The Exposed City

There is a vast amount of information about a city which is invisible to the human eye – crime levels, transportation patterns, cellphone use and air quality to name just a few. If a city was able to be defined by these characteristics, what form would it take? How could it be mapped?

Nadia Amoroso tackles these questions by taking statistical urban data and exploring how they could be transformed into innovative new maps. The "unseen" elements of the city are examined in groundbreaking images throughout the book, which are complemented by interviews with Winy Maas and James Corner, comments by Richard Saul Wurman, and sections by the SENSEable City Lab group and Mark Aubin, co-founder of Google Earth.

Nadia Amoroso specializes in visual representation as it relates to architecture, landscape architecture and the urban environment. She is a Lecturer at the University of Toronto, John H. Daniels Faculty of Architecture, Landscape, and Design. Amoroso was also a recent Visiting Fellow at Cornell University and Visiting Professor at the University of Arkansas. She has a Ph.D. in Architectural Studies, and degrees in Urban Design and Landscape Architecture. Amoroso is a principal of Orange + Blue Consulting, specializing in urban design, mapping and visual representation (digital and traditional means).

The Exposed City

Mapping the urban invisibles

Nadia Amoroso

LONDON AND NEW YORK

First edition published 2010
by Routledge
2 Park Square, Milton Park, Abingdon, Oxon, OX14 4RN

Simultaneously published in the USA and Canada
by Routledge
270 Madison Avenue, New York, NY 10016

Routledge is an imprint of the Taylor & Francis Group, an informa business

© 2010 Nadia Amoroso

Typeset in Univers by Keystroke, Tettenhall, Wolverhampton
Printed and bound by Grafos SA, Barcelona, Spain

All rights reserved. No part of this book may be reprinted or reproduced
or utilised in any form or by any electronic, mechanical, or other means,
now known or hereafter invented, including photocopying and recording,
or in any information storage or retrieval system, without permission in
writing from the publishers.

Every effort has been made to contact and acknowledge copyright
owners, but the authors and publisher would be pleased to have any
errors or omissions brought to their attention so that corrections may
be published at a later printing.

British Library Cataloguing in Publication Data
A catalogue record for this book is available from the British Library

Library of Congress Cataloging-in-Publication Data
Amoroso, Nadia.
The exposed city : mapping the urban invisibles / Nadia Amoroso. – 1st ed.
p. cm.
Includes bibliographical references and index.
1. Urban geography–Graphic methods. I. Title.
GF125.A46 2010
307.76022'3–dc22
2009039717

ISBN10: 0–415–55179–X (hbk)
ISBN10: 0–415–55180–3 (pbk)
ISBN10: 0–203–85537–X (ebk)

ISBN13: 978–0–415–55179–3 (hbk)
ISBN13: 978–0–415–55180–9 (pbk)
ISBN13: 978–0–203–85537–9 (ebk)

Contents

Foreword by Richard Saul Wurman — vii
Preface — xi
Acknowledgments — xv

Part I Essays — 1

1 Map or Drawing? The Visual Expressions of Hugh Ferriss — 3

2 Graphic Integrity and Mapping Complexity: The Works of Lynch, Wurman and Tufte — 41

3 The "Datascapes": The Works of MVRDV — 68

4 The Map-Art: The Works of James Corner — 93

Part II Visuals — 115

5 Drawings: The Map-Landscapes — 117

Afterword Look Ahead Comments by SENSEable City Lab and Google Earth Inc. — 159

Bibliography — 168
Index — 174

Foreword

Richard Saul Wurman

It all starts with a line in the sandy earth with a stick.

A line in time with a route marked with notches, demarcating distance, observed landmarks, time and destination.

Sometimes it starts with a question, Where am I? What's around me? How do I get there? How can I understand the relationship between usage and time, geology, demographics, meteorology, pollution, pattern after pattern?

Sometimes it's a view that you can't see or measure, as in *Civitates Orbis Terrarum* or it's a city plan, which has all the public buildings shown at their ground floor level with columns and courtyards, piazzas and the streets all rendered. All the places that are private are solid, as in *Nolli's* map of Rome. Perhaps, it's Paris from the eye of a bird before flying was possible, even before hot air balloons were a dream.

It's Man's Ability to Perceive, it's the MAP. It's also the map through time with the ease of quick time and computer graphics and morphing, changing one pattern into another. Time telling a story through a day, a week or a year. Time showing change, it's the transparency of information combined with other information creating a third piece of information.

It's everybody and I mean everybody seamlessly using Google Earth, performing the magic of flying around the earth up and down and into somebody's back yard and with special programs and insets seeing the animals marching through Africa.

It's Marvin Minsky's *Mapping the Mind*, it's the mind's eye. It's the pattern of orientation.

It sometimes ends with silly questions: Is it within walking distance? And getting a verbal explanation with three directions too much results in having to ask and ask again.

Or it's a whole city like Tokyo, where asking questions to find a place is part of the culture, because the numbering of buildings is not sequential but based on the order of their construction.

Nadia Amoroso has given an extremely thoughtful definition of the idea of MAP with broad slices from Ferriss to Tufte to the *datascape* concept – that define the geographic and artistic definitions of MAP from drawing, from the visual expressions to be found in drawings, their diagramed analysis and statistical representations. The most exciting new characteristic of the new maps are statistics through time, the visualization of changing complex data that allow one to see the things they've always seen but never seen and certainly with an ease of visual representation. She has brought to the forefront a visually exciting way to view information that has spatial, creative and suggestive outcomes. With a background in landscape architecture, urban design and in architecture, she crafts maps as some kind of landscape form perceived by the data type. Maps as metaphor, the data as metaphor, map as art, data as art.

Muriel Cooper from MIT's Media Lab, in what became her final major presentation before an open-mouthed TED (Technology, Entertainment, Design) audience and just months before her death, gave a predictive hint of the effortless ability that we were to have soon, to fly through information.

That peephole into the future was the future, is now our present. I create maps in order to understand something. . . . it is a journey.

Amoroso exposes the invisible elements of the city through 'map-landscapes' – the theory has now been embraced, the map has now been embraced, as the preeminent way of organizing and understanding complex data relative to demographics, marketing, the environment, traffic patterns, as well as the less romantic descriptions of crime and unrest.

In the early times of exploration the map itself had power and therefore value as its possession allowed knowledgeable access to information. Once again the map has power and value. As the explosion of visualization of information both cartographic and statistical becomes commonplace, it will have parallel discoveries in the mapping of the human body, its health and disease.

Maps have been powerful, maps have been beautiful, maps have been dense with all manner of information, coated with flora, fauna, heights and depths, land use, people use, traffic flows and particularly hostile borders.

I bring up the last as a reference to the official UN map, which is a map of the globe with lines, often white, separating countries, emphasizing the separation with that notation, which is really the notation of hostility between nations, walls are built, customs stations erected, wars based upon that line begin to annex property across that very line, determine ownership of water, sometimes religion and often wealth. Getting a green card is dependent on that white line.

Historically, maps have determined a lot and continue to do so. Maps, atlases, all the above, have had a long history of annotating data in a clearer and more exact manner. The major effect of technology has been accuracy, access to cartography, as in Google Earth, formerly called Keyhole, which has popularized the map – made even clearer the "everything takes place some place" idea.

We are on a cusp and Nadia Amoroso is one of the team players on this cusp of the New Map, often animated, showing patterns with great clarity and singularity, combining scientific, physical structures, atmospheric conditions and showing these patterns over time, a day, a week, a month, a year or a decade. The focus is also interesting because the focus is basically on where people live on an urbanized globe.

I've referenced the globe of countries; Buckminster Fuller talked of our blue globe, the globe of water, which covers approximately 72% of our spheroid.

Nadia Amoroso, as are others, is focusing on understanding with clarity and over time, a planet earth for people and an urban globe.

Preface

This book examines mapping visualization as it relates to visual representation of the city from the early twentieth century to today. It profiles the contemporary history and theory of mapping explorations in terms of exposing the "invisible" forces that shape our urban environment. Specifically, it reviews the role of maps as both presentational results as aesthetic objects and informative tools, which could be used to influence architectural, landscape architectural and urban design moves.

How can the abstract forces shaping urban life be rendered artistically, spatially and informatively in the form of alternative "maps" which represent the urban invisibles, and are not usually accessible to routine professional expertise (such as the urban designer or architect) or the ordinary urban dweller? In what ways do the new visualizations provide deeper insights into the city than conventional maps are capable of revealing? We will examine mapping representations and methods that reveal the "unacknowledged of the city," reviewing the contemporary history and theory of mapping explorations in terms of exposing the "invisible" forces that shape our urban environment.

Maps provide layers of geographical data that direct one through the city. They guide the user in space and reveal specific parameters of the city. The term "map" is defined and redefined throughout the research and is supported by a series of visual experiments that lead to alternative depictions of the city as hybrids of poetically striking maps and images for global cities. Imagine a city invisible to the human eye and only manifested by its non-visual urban phenomena. What shape will it take? With the new urban forms represented as images, would they become guides into the city – in effect spatial maps? Revealing forces not visible to the naked eye, the work abstracts textual urban data and transforms them into architectural visions. Examples of these forces as depicted today in the urban context would include criminal activities, zoning by-laws, population densities, transportation patterns, public surveillance, cellphone usage, air-quality readings, and other spatial statistics. What are the invisible mappings of the future?

Using new modes of representation, a series of multidimensional maps will guide the viewer into alternative readings of the city as both artistic and informative artifacts. These maps express otherwise unseen urban data and bring them to the forefront – *"exposing" the city*. An image of the city that has artistic merits attracts its viewers with its seductive qualities and is likely to inspire immediate reactions – which can be said to bridge the gap between visual art and urban design. Gauging the extent to which a scheme is capable of manifesting those invisible urban forces which define the essence of a city also determines the usability and value of its mode of representation.

We begin by examining a series of key individuals from the early to late twentieth century whose works can be analyzed as precedents for the study of maps as both informative and aesthetic productions that have provided positive outcomes. The first chapter reviews the powerful renderings of the early twentieth-century architect, Hugh Ferriss, specifically his 'Evolution of the Set-Back Building' drawings in relation to the zoning Ordinance of 1916, and questions the role of these drawings as three-dimensional maps and as products in revealing the hidden factors of New York City. Perhaps Ferriss's New York City Ordinance drawings are the first true "datascapes" and three-dimensional map-drawings. He refines the "definition of map," combining a perfect balance between the logic (data) and the presentation (art). The mapping conventions developed by Kevin Lynch from his seminal book *The Image of the City* (1960) will then be discussed. His maps exposed the physical elements of the city. Lynch uses qualitative graphics to define qualitative information (visual data) – the physical elements of the city. Information architect guru, Richard Saul Wurman first pushed the notion of information graphics and mapping, beginning in the late 1960s. Wurman, by training, is an architect who studied under Louis Kahn. Wurman's lifelong quest is to make "visual" things that he does not understand. In other words, he creates "maps" to help guide him along a journey of discovery and conclusion. He also set out a series of principles that he believes guides him in the creation of a "powerful" map. He continues to do this today in his 19.20.21 project, which examines 19 cities in the world with over 20 million in population in the 21st century. Information graphics expert Edward Tufte set forth guiding principles to achieve graphical integrity in order to provide truthful communicative visual measures. In the late 1990s and turn of the millennium, Dutch architectural firm MVRDV conceived of the "datascape" concept and provided us with provocative mapping visualizations as potential urban designs, using data as the palette for form. MVRDV used qualitative (visual approaches) to signify quantitative (numerical data) environmental problems data. They created uncanny, shocking graphics of "what-if" situations for the Netherlands. There are many groups and individuals rising to the challenge of developing maps that are not just the typical GIS outcomes, but rather poetic visual documents that are telling of a new city or landscape. Contemporary landscape architect and academic James Corner utilizes mapping as a creative and inventive process and shows the potential of maps to unfold "hidden" facts of the city. Today, the SENSEable City Lab at MIT tries to push the envelope in developing alternative spatial representations of data that expose the hidden factors of the city. Almost all design professions concerned with site and urban development are using Google Earth, and each year the Google Earth team aspires to develop powerful layers that expose new elements of the city – such as their recent historical maps – and ocean layers that allow you to explore the other 70 percent of the world. Today, thanks to the efforts of some courageous and forward-thinking individuals, architects and urban designers are more receptive than ever before to using alternative modes of representation to unravel the cityscape's *deeper* issues. These individuals revealed certain aspects of the city through forward-thinking artistic, spatial and informative mapping approaches.

We will conclude with a series of "visuals," defining an artistic and technical approach towards the contemporary practices of urban representation – a series of experiments undertaken to develop new digital visualizations – or "mappings." The new maps are intended as an alternative to the mainstream spatial representations of the contemporary city – a re-visioning and rethinking of mapping and urbanism for the twenty-first century, as well as an exercise in representing the hidden city through visualization and animation of architectural and urban data. It will profile three global test cities – Toronto, New York, and London.

When is something a work of art and when is it merely statistics? This is a topic of investigation throughout the book. What visual means are necessary to convey both artistic and informative properties as they relate to mapping – representations that create a balance between the value of the presentational results as aesthetic objects (the art, the creativity) and their empirical evidence (the data, the logic)? These experimental mapping works suspend these two realms, and merge them – creating a marriage between the expressed and the unexpressed, between fact and fiction, between idealisms of an ordered and perfect future and practicalities of design. The experimental mapping works explore the possibility of creating, like Ferriss and the many other individuals profiled, new kinds of urban form drawn by urban data, generated and displayed digitally. This experimental modeling is pushing the GIS analysis into a new era. Cities are storing and sharing data electronically these days which now enables this kind of experimental mapping to happen more readily.

These images can be defined as "*map-landscapes*" – prefiguring some kind of landscape form or spatial outcome based on data. Data such as crime rate indices, traffic flux, density, public surveillance, market values and other textual information relating to cities are visualized into multidimensional artifacts that create tantalizing new form of the city, yielding better design initiatives. Maps as artifacts, guides, art, spatial form, images, design tools, metaphors, and landscapes – these functions are means to provide deeper insights into the city while attracting the urban planner, architect, landscape architect, designer and the common urban dweller. These roles could in fact provide stronger communicative forces that can influence planning and design decisions. Throughout the process of transforming data into landscape form, an investigation of the actual and possible creation of new urban form is transcribed. The images are composed within a field set (set of parameters) in which the quantifiable data are examined and analyzed for visual abstraction. The work transforms a series of urban realities into provocative, abstract urban forms. These "map-landscapes" are intended as a set of instructions or perhaps tools for audiences such as the urban designer, the landscape architect and the architect, and are meant more as an immediate reaction for the urban dweller. The representations are created through a digital process taking a flat surface of embedded information, and tapering and folding this surface. As a flexible mesh, the surface of the map is active and crafted, based on the parameters of the value set of the data, which then takes a final form. The generated "form" is considered both a multi-dimensional map and a new spatial outcome for prefigured designs, qualifying them as "map-landscapes."

Acknowledgments

This book would not be possible without the effort and support of key individuals.

I would like to acknowledge many of my colleagues who generously offered their time for reading sections of early manuscripts and providing vital feedback, especially James Patterson from the University of Oklahoma; Jeff Shannon, John Crone and Fran Beatty from the University of Arkansas; Peter Trowbridge from Cornell University; and Ina Elias from the University of Toronto.

I am grateful to Philip Steadman and Penelope Haralambidou from the Bartlett School of Architecture and Graduate Studies who have shared their knowledge on this subject matter and have guided me to further explore this topic. I would like to thank Alan Penn from the University College London and Frank Duffy from DEGW for their thoughtful insights and advice on this subject.

I would like to express a special thanks to Richard Saul Wurman for writing the book's foreword and for his charismatic attitude throughout this process; and to Winy Maas (MVRDV) for his generous time and for offering his thoughts on the role of datascapes and their future applications; and to landscape architect James Corner (the University of Pennsylvania/Field Operations) for commenting on the role of artistic mapping representation and the relationship between scale and graphic type. I would also like to thank Mark Aubin (Google Inc./Google Earth) and a number of key members from MIT's SENSEable City Lab for providing some concluding thoughts on advanced digital technologies for representing the city. I would like to thank many of my students from my digital media course and my mapping visualization course for providing some of the material presented in this publication. I want to thank my former student, Tara Razavi, for her assistance. I also want to thank my publisher Routledge (Taylor & Francis Group) and the editorial vision by Alex Hollingsworth and Louise Fox for bringing this publication forward.

I would like thank my colleagues from John H. Daniels Faculty of Architecture, Landscape and Design at the University of Toronto who introduced me to this subject matter many years ago, especially Professors R. el-Khoury and D. Lieberman.

Finally, I am grateful to my family – my parents, sister, James, Carlo, Karen, Tina, Joseph and little Siena for their ongoing support and encouragement; and to my husband, Haim, for his devotion and patience, which has made this publication process a rewarding experience.

Nadia Amoroso, Ph.D.

Part I

Essays

1 Map or Drawing?

The Visual Expressions of Hugh Ferriss

Introduction

This chapter focuses on the role of drawings as graphic sources for investigating visually intangible conditions within the urban realm. The capacity of drawings to reveal possible interpretations of the city as visually and sometimes emotionally accessible information qualifies them as maps. Revealing a multidimensional view of their subject, maps are encoded with many layers of technical and abstract data reflecting the legal, environmental, economic, social and political circumstances within a city. Through the manipulation of drawing conventions and the use of abstract signs, maps guide their viewer through a maze of an artificially constructed field of forces which define the physical reality of the built environment. However, maps can never be understood as purely objective representations. While the making of a map often involves a lengthy process of gathering and interpolating large quantities of statistical materials, maps are highly controversial artifacts, which register the prevailing political demands of their cultural context and the personal input of their makers.[1]

    As products of a thorough investigation of a wide range of factors, maps produce both a precise and an imaginary synthesis of the present and at times future conditions within a city. If rendered by the hands of a talented artist, maps also become seductive artifacts which attract their audience with their imaginative and graphic qualities. As such, maps not only communicate a possible objective reality, but are also charged with the emotional input of their artist-maker; they embody human dimensions and experiences. They become particularly revealing evidence of urban circumstances as they are perceived and felt by individual inhabitants. Described as simultaneously artistic and informative artifacts, maps mirror and graph both the complexity of the factors that define the present reality of an urban situation, and the intuitive processes which led to their creation. As sources through which predictions about the future of a city can be made, maps are essential tools in the professions of urban planners and architects.

Hugh Ferriss: the art of mapping the invisibles

Focusing on the works of early twentieth-century artist and architect Hugh Ferriss (1889–1962), with a particular emphasis on his graphic interpretations of the 1916 zoning Ordinance of New York City, this chapter explores the potentials of drawings as media through which "invisible" dimensions of cities can be explored and revealed. Ferriss's depictions of the zoning laws, in drawings of the *Evolution of the Set-back Building*, can be considered three-dimensional maps, and arguably indispensable for understanding the architecture of early twentieth-century America. Not only did his drawings become expressive vistas into the future of Manhattan's architectural and urban design conditions, as legacies of one of the most talented artists of the period, Ferriss's depictions also synthesize the positivist and progressive spirit of their era. These drawings foreshadowed a city which, due to its threatening qualities, was destined to remain as only pictorial. They became visual guides of the spatial container in which architects and planners can build, and these drawings helped clarify the legal and textual confusions of the by-laws. Nevertheless, his drawings left a lasting impression on their contemporary and later audiences, while their fantastic and prophetic qualities establish them as iconic evidence of a visionary brand of architecture on a par with the works of some of the greatest visionaries of all time such as Piranesi and Boullée:[2]

> Ferriss was, in a way, an apostle of bigness – stimulated by the sight and feel of the mighty construction efforts going on around him, he exaggerated their scale in his drawings. Thus he has appropriately been compared to Piranesi, with the crucial difference that the vision of that 18th-century Italian artist, based upon the monumental relics of a long-gone past, was intensely ambiguous when not outright gloomy about human prospects. Ferriss, by contrast, extrapolated a vision of a (to him) brilliant future from the ever-changing cityscape of the present.[3]

The striking character of Ferriss's drawings established the reputation of the artist as one of the most skilled draftsmen of his time. Ensuring him great success in his career, his works captured the imagination of a large audience. The extensive publicity that Ferriss received arguably affected the thinking of his contemporaries and provoked responses from them. This is one of the reasons why his drawings/ maps were so engaging. They drew the attention of a wide audience: the architect who was concerned with the overall built form; the urban planner who was interested in overall urban massing pattern; the developer who was interested in the overall floor area ratio and economical rentable space; and the urban dweller and city officials who were interested in the vitality of the city and allowing more light and air onto the streets of New York City.

In an attempt to understand the ways in which the graphic works of Ferriss functioned to inform and influence the perception of their viewers, we look at his *Evolution of the Set-back Building* drawings in relation to the zoning Ordinance of 1916, as both works of art and maps that exposed a "new" New York City.

The drawings

The drawings of Ferriss contributed to an all-encompassing awareness of the effects of zoning and its architectural and urban design implications. While his drawings allowed architects to sharpen their vision of a new style, commensurate with the technical innovations of their time and inspired by the aesthetic possibilities of zoning, they signaled to socially minded individuals the menacing conditions that could arise in the future. Presenting Ferriss's maps as catalysts for the innovations which characterize the architectural movements of the early 1920s may seem an exaggerated proposition. However, the fact that his drawings consolidated and provided a clear definition of the major legal and economic shifts within the milieu of early twentieth-century American culture cannot be denied. His images spoke to all sets of audiences simply by their familiar architectural form and style. This was important, because as works of art, these drawings spoke to the general public – the citizens of New York. As maps, they guided the city planner and architect into the unknown of the zoning Ordinance. Contributing to an assessment of zoning and its consequences, Ferriss's drawings are best described as one of the significant agents which illuminated and prepared the masses for the great changes that were about to take effect. Surfacing at a critical time, they highlighted and strengthened the prevalent mood of their age.

The influence of Ferriss on his audiences was also a consequence of his artistic abilities. In order to understand the precise methods by which the artist transformed the textual content of the Ordinance into highly expressive charcoal renderings, a close inspection of his drawing techniques and use of medium will conclude our investigation. In addition to a study of the drawings of the zoning Ordinance, much can be learnt from the writings of the artist on the art of rendering. A spokesperson for the rising profession of an "architectural delineator," Ferriss lectured on the subject of rendering while regularly contributing articles to various journals. These materials are invaluable sources through which we gain a better understanding of the artist's techniques. They provide us with first-hand clues about the ways in which he bridged the gap between text and image, while informing us about the methods he used to imbue his works with such gravity that it was impossible for the popular press, architects and planners to ignore them.

The power of the map-drawing

A powerful map embodies four characteristics: it is *informative*, *revelatory*, *seductive*, and *suggestive*. These categories are highlighted in the works of Ferriss to give special potency and credibility to his renditions of a contemporary and future Manhattan. In his drawings these characteristics combine to attract attention and influence perceptions of the urban environment.

Both Ferriss's artistic and textual productions place him deeply within the debates among architects and urban planners who in the early 1920s tried to anticipate and come to terms with the new zoning law as it related to their professions. Consequently, his works cannot be understood without a consideration of the variety of forces which first led to the formulation and later application of the parameters of zoning in actual practice. Understanding the status of architecture at the time the zoning Ordinance was introduced, the coming together of the intrinsically opposing economic and reformist ideologies which facilitated the introduction of zoning, and the attitudes of architects, urban planners and the general public to these changes are all topics which will inform and enrich our investigation. The summation of these factors can be identified as the "invisible" forces which are compounded together and are given concrete shape in the drawings of Ferriss. The harmony of these forces within his drawings infuse them with visual power and potency, for what they encapsulate is an infinitely more complex depiction of urban and social facts, relevant to an understanding of early twentieth-century American life, than what his pictures would show had he only intended to produce mere literal translations of the legal material. A virtual city was implied by the zoning Ordinance; Ferriss exposed this city as it reflected the restrictions on form, the economic demands of property owners and developers, and the aesthetic concerns of architects.

Architectural art critic Christopher Humes says: "it was architectural renderer Hugh Ferriss who captured the romance of the highrise. His darkly evocative drawings pointed to the esthetic potential of this quintessential 20th-century building type."[4]

The 1916 New York Ordinance

Drafted primarily as a response to the unsteady rate with which high-rises were spreading through downtown Manhattan, the zoning Ordinance was enacted in 1916 by the New York City Board of Estimates and Apportionment as the first comprehensive zoning law in the country.[5] Two men were responsible for writing the Ordinance: architect George Ford and statistician Robert Whitton. A review of the historical events which led to the approval of the zoning laws by city officials proves that a range of social, political and economic agendas lay behind the actions of individuals who finally conceded the values in zoning. In the early 1900s, the high value of land resulting from the concentration of businesses on the lower tip of

Manhattan Island together with advance in steel-frame construction established skyscrapers as a highly profitable office building type.[6] Realizing the economic merits of tall buildings, developers and landowners in an attempt to maximize returns on their investments built monstrous structures into which they crammed as much space as their budgets and the buildings' sites permitted.[7] City officials conceived zoning as a set of legal measures to combat the irresponsible development of real estate, which affected the built environment by causing dark and congested conditions in the city.[8]

As the first comprehensive zoning laws in the country, the 1916 Ordinance subjected every piece of real estate in Greater New York City to three types of regulation: *use, height, and floor area limitations*.[9] However, in high density commercial zones, in which most office towers were found, zoning primarily affected the height and volume of buildings.

Ferriss's visual response

In 1922 Ferriss illustrated in a series of four drawings the effects of zoning on the design of a typical skyscraper. Referred to as the *Evolution of the Set-back Building* or more commonly as the "Four-Stages" drawings, these illustrations, which first appeared in an article in *The New York Times* magazine in March of the same year, represent one of the earliest attempts at studying the formal consequences of the zoning laws. The laws had implied an "invisible" city; the drawings were intended to yield a "visual proof" of this. The drawings address sections of the Ordinance which explain restrictions on the height and volume of buildings.[10] While the drawings prove the artist's thorough understanding of the terms of the Ordinance, they also attest to his awareness of the overriding economic forces which were so fully intertwined with the practice of architecture and urban design in the early 1900s. An examination of the drawings in relation to the requirements of the zoning guidelines reveals the process by which the artist translates legal codes into visually powerful documents. Ferriss's trademark style of rendering, using black charcoal to create a heavy chiaroscuro, gives his works special graphic power. The expert handling of the drawing media conferred on Ferriss's drawings an appealing aura which resulted in great publicity and attracted a wide range of audiences that included architects, urban planners, and arguably developers, who were familiar with his work through advertisements.

"Stage 1" map-drawing

The images depict the imaginary envelope of a building for a full city block in Manhattan. In a hypothetical site, measuring 200 × 600 feet, a building mass, reflecting all the requirements of zoning, evolves through a series of transformations to emerge as a close representation of an architectural form that is considered

1.1
Ferriss's drawing from the *Evolution of the Set-back Building Series*: "Stage 1." The first drawing of the series represents the maximum permissible building mass under the zoning laws. As a simple expression of mass, the image describes a composition of bulky, pyramidal forms arranged in an elegant manner. Ferriss captures the poetics of the data through his chiaroscuro techniques.

Source: Hugh Ferriss, *The Metropolis of Tomorrow*, 1986, p. 73

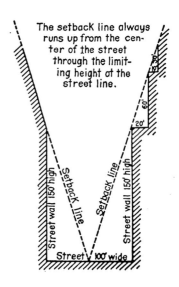

1.2
"Setback Principle Diagram" used by Corbett. Diagram showing setback principle in a "one and one-half times" district for a 100-foot-wide street.

Source: Carol Willis, "A 3-D CBD: How the 1916 Zoning Law Shaped Manhattan's Central Business District." *Planning and Zoning New York City: Yesterday, Today and Tomorrow.* Ed. Todd W. Bressi, 1993, p. 15

aesthetically, legally and economically. The first drawing of the series represents the maximum permissible building mass under the zoning laws. As a simple expression of mass, the image describes a composition of bulky, pyramidal forms arranged in an elegant manner (Figure 1.1). In order for this drawing to delineate a truthful interpretation of the Ordinance, two important numerical values had to be calculated. The first is the height of the first set-back, which can be seen in the drawing as the point about which the building envelope begins to taper, and the second is the area for the base of the tower.

The height of the first set-back, which determined the dimensions of the maximum building envelope, was based on a formula in which the width of the street and a value corresponding to the building's zoning district were the two significant coefficients.[11] While the width of the street was in most cases fixed by existing urban conditions, reflecting the restrictions of a site and the size of lot, the district variable was introduced as a parameter that specifically accommodated the building's height and volume to the density of the neighborhood within which it was located. The purpose for the latter component was to enable officials to control the city's growth by stopping the erection of high-rises in certain parts of Manhattan. The procedure for calculating the height of the first set-back and the maximum volume involved marking the center of the street and projecting from it a diagonal line (describing a plane in two-dimensions). The point at which this line hit the building's profile, as it was vertically extruded from the lot line, established the height of the first set-back. An example of this process can be seen illustrated in a diagram which was used by Harvey W. Corbett (1873–1954) at the fifty-fourth annual convention of the American Institute of Architects to explain the effects of the laws (see Figure 1.2). Depicted in this image are the conditions for a 1:1½ district for a street that is 100 feet wide.

Assuming a relatively dense context, Ferriss sites his building within a 1:2 zoning district. The angle of the sloping planes in the first drawing result from the calculations according to the technique describe above. Ferriss reiterates the zoning requirements in his 1922 article:

> The height of this wall is based on the width of the street which it faces. There are five in New York of varying height allowances. In the tallest a building may rise two and one-half times the width of the street before it begins to slant backwards; in the lowest only once the width of the street.[12]

An amendment to the zoning laws allowed a tower of unlimited height as long as it occupied in plan only 25 percent of the lot. The first of Ferriss's drawings exhibits the two basic requirements explained above. While the sloped planes reflect the zoning set-back requirements, a tower of indefinite height has only been cut short by the limits of the piece of paper.

The legislation of the 1916 Ordinance had equipped city officials with the technical tools and legal authority they needed in order to improve living conditions for the urban dweller. However, the purely practical basis of the Ordinance had not permitted its writers to consider the aesthetic consequences of the laws. The basic rules of set-backs only carried a design so far, and as Ferriss had warned his viewers, the final resolution of the masses in a building (as in his fourth drawing) was not intended as the last step in its development. In an effort to emphasize the architect's role within the design process, both Ferriss and Corbett continued their collaboration to produce two images in which each person separately expressed his concept of what a fully articulated building could look like (Figures 1.3 and 1.4). While both images were executed by Ferriss, the differences between the treatment of architectural elements and detailing within each design show two intrinsically diverse mentalities. Corbett imagined his building as an eclectic collage of modern and historic styles; Ferriss's solution gained its architectural thrust through its sculptural massing and almost total lack of decorative detailing. His vision came to dominate

1.3 and 1.4
Corbett and Ferriss sketches: "Final Stage, Possible Developments within the Envelope." Ferriss and Corbett continued their collaboration to produce two images in which each person separately expressed his concept of what a fully articulated building could look like. While both images were executed by Ferriss, the differences between the treatment of architectural elements and detailing within each design show two intrinsically diverse mentalities. Corbett imagined his building as an eclectic collage of modern and historic styles; Ferriss's solution gained its architectural force through its sculptural massing and almost total lack of decorative detailing.

Sources: (1.3) Harvey Corbett, "Zoning and the Envelope of the Building," *Pencil Points*, 4, April 1923, p. 14. (1.4) Ferriss, *The Metropolis of Tomorrow*, 1986, p. 105

the architectural trends of the period, while his technical mastery and foresight inspired many architects, such as Raymond Hood with whom he was certainly a popular figure.

Two important documents from the early 1920s can be identified as amongst the earliest attempts at illustrating the architectural consequences of the zoning laws. The earlier of the two is a series of diagrams used by Corbett in a presentation at the fifty-fourth annual convention of the American Institute of Architects (AIA). Shortly after this event, Corbett's lecture was published as an article entitled "High Buildings on Narrow Streets," and can be cited as one of the first discussions of the aesthetic possibilities of zoning.[13] The publication of the *Evolution* drawings by Ferriss in his 1922 article represents the second source.[14] The simple line drawings which Corbett uses to support his arguments were originally made by New York architect George

1.5

Modest Zoning Diagrams developed by George Ford and the Commission on Building Restrictions, New York, 1916. George Ford, an early twentieth-century architect and one of the authors of the 1916 zoning Ordinance, used these diagrams to communicate the transformations of the permissible bulk of a building according to its site conditions.

Source: Carol Willis, "Drawing Toward Metropolis" in Ferriss's *The Metropolis of Tomorrow*, 1986, p. 157

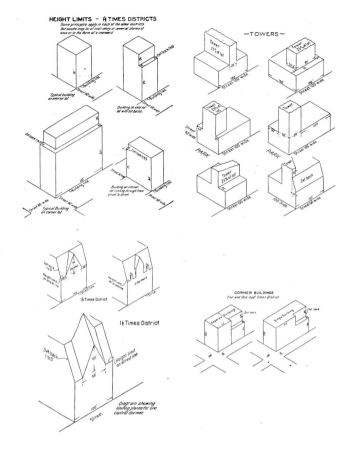

Ford, who was one of the individuals responsible for writing the content of the Ordinance. Through a series of axonometric views and one diagrammatic section cut through a street within a 1½ district, Ford intended to communicate the transformations of the permissible bulk of a building according to its site conditions (Figures 1.2 and 1.5). While the sloped surface of the building envelope on the left side of the street in the section drawing is a direct reflection of its maximum bulk, on the right side it is architecturally resolved through a series of stepping terraces. Ford's drawings, however, only projected an incomplete, misleading interpretation of zoning.[15] Not only do the simple combinations of cubic buildings and towers run contrary to the aesthetic aspirations of the period, the buildings as represented in his diagrams are drawn to fit small lots which characterize the pre-zoning conditions of architecture in New York.[16] This was evident as Corbett, soon after his talk at the AIA convention, commissioned Ferriss to produce his illustrations. The much improved and thorough resolution of the zoning criteria as represented in the *Evolution* drawings can be explained as interventions by the artist.[17] Ferriss's renditions are of exceedingly superior quality, both in their graphic and informative content. This is supported by Willis: "The iconic quality of Ferriss's images contrasted starkly with the black line diagrams prepared by George B. Ford for the committee"[18] (Figure 1.5).

Map or drawing?

These modest diagrams were the only developed set of visual guidelines before Ferriss's drawings. Willis continues to argue the importance of Ferriss's zoning drawings as opposed to Ford's diagrams in the following words:

> [Ford's diagrams] presented about twenty analytical figures, which outlined the maximum building envelope for a variety of small sites (corner, mid-block, etc.) in different zones. Though the diagonal planes of the zoning envelope were generally squared off to suggest cubic structures, some diagrams included dormers, and the towers were truncated after several stories. Ferriss's compelling images, on the other hand, offered architects directions and inspiration. Not only did he clarify how the formula could be profitably developed in a large, steel-skeleton structure, he also made the buildings look like Architecture, revealing the latent beauty and power of the simple setback form.[19]

The sketches shown in Figures 1.2 and 1.5 are definitely not as compelling as Ferriss's charcoal renditions. They provided the spatial container in which the architect or developer can build. In essence, the drawings became spatial guides that prefigured the future Manhattan – the three-dimensional maps. Much like the architects of his time, Ferriss was intrigued by the aesthetic potentials of zoning. Ferriss celebrates the detachment of zoning from any aesthetic criteria, and instead sees the value of the laws in their practical thrust. In the introduction to his illustrations of the Ordinance, he recognizes the practical genesis of zoning and goes on to consider their fundamental impact on the practice of architecture. As he puts it, "the recent enactment of the New York zoning laws creates a unique situation – 'restricting' in nature, they are producing a profound evolution in architecture."[20] A whole new understanding of architecture was made possible through the introduction of zoning. Ferriss explains in his famous book, *The Metropolis of Tomorrow*:

> the actual effect of the law was to introduce what is often spoken of as no less than a new era in American Architecture. The whole procedure constitutes another example of the fact that the larger movements of architecture occur not as the result of some individual designer's stimulus but in response to some practical general conditions.[21]

"Stage 2" map-drawing

Representing the second evolutionary stage of the set-back building is a drawing in which a volumetric expression, similar to the definition of mass in the first stage, is further articulated with a series of vertical cuts arranged rhythmically along the base and the shaft of the tower (Figure 1.6). Identified as "light-courts," these incisions in the bulk are primarily intended to relieve dark interior spaces by allowing the penetration of daylight. The introduction of light-courts into the building mass according

1.6
Ferriss's drawing from the *Evolution of the Set-back Building* Series: "Stage 2." Hugh Ferriss depicts the second evolutionary stage of the set-back building. The drawing shows how a volumetric expression, similar to the definition of mass in the first stage, is further articulated with a series of vertical cuts arranged rhythmically along the base and the shaft of the tower.

Source: Ferriss, *The Metropolis of Tomorrow*, 1986, p. 75

1.7
Ferriss's drawing from the *Evolution of the Set-back Building* Series: "Stage 3." Hugh Ferriss depicts the third evolutionary stage of the set-back building. The building envelope attains an appearance that is both architecturally and economically practicable with its steps.

Source: Ferriss, *The Metropolis of Tomorrow*, 1986, p. 77

to Ferriss, marks the first intervention by an architect.[22] The mass as it is represented in the second-stage drawing is still a far cry from a fully developed building. The angled planes which dominate a considerable portion of the building envelope reflect a literal interpretation of the set-back requirements, while the infinitely high tower, bleeding off the page, responds to that section of the zoning Ordinance which, as previously mentioned, allowed construction to an unlimited height on 25 percent of the lot.

"Stage 3" map-drawing

In the third stage, we get another guide into the city block. It is at the third stage of its evolution that the building envelope attains an appearance that is both architecturally and economically practicable (Figure 1.7). While the artist recognizes the imaginative possibilities of the sloping planes as "stirring a certain type of mind," in the third drawing of the series he exposes a practical solution for transforming the angled planes. Portraying the building mass as a convincing architectural and economic proposition, rectangular stepped forms replace the sloped surfaces. The new formation suggests a more conventional approach to architecture in which the vertical walls also define habitable spaces in the interior. The schematic character of the building at this stage is particularly evident at the tips of the two blocks (masses), flanking the central tower, in which a multitude of small steps gradually terminate the pyramidal masses.

"Stage 4" map-drawing

The fourth and final drawing that Ferriss produces to explain the formal consequences of the zoning laws is a close representation of a building with no exterior articulation (i.e. windows, decoration). In this drawing (Figure 1.8), the built form is portrayed as a sculptural impression of a completed design for a skyscraper in which the careful arrangement of the masses in accordance with the legal and economic measures of zoning ensures the building's practicability. However, Ferriss concludes with one warning: "This is not intended as a finished and habitable building; it still awaits articulation at the hands of the individual designer."[23] Here he portrays a building, which alludes to a built form without articulation.

Practical and economic considerations lay at the core of the Ordinance. Through both text and image, Ferriss skillfully responds to and interprets the zoning criteria. While the first two drawings of the *Evolution* series mainly address the legal requirements of the Ordinance, designed to alleviate congestion and increase light and air in the city, the last two drawings assimilate the architectural consequences of the zoning laws as they reflect the financially established conditions within a real-estate market. Conceived as visual proofs of formerly intangible facts about the city and its architecture, Ferriss's drawings testify to the economic implications of zoning and its manifestations in architecture. While the introduction of

1.8
Ferriss's drawing from the *Evolution of the Set-back Building* Series: "Stage 4." The fourth and final drawing that Ferriss creates to demonstrate the formal consequences of the zoning laws is a close representation of a building but with no exterior articulation.

Source: Ferriss, *The Metropolis of Tomorrow*, 1986, p. 78

zoning had threatened the financial interests of developers and property-owners, architects were dedicated to provide their clients with the best service.

From text to drawing to guide

An analysis of Ferriss's artistic and textual productions reveals that he was cognizant of and responsive to debates within the architectural community.[24] As a visionary, Ferriss not only recognized the complications which zoning entailed for architects, but proceeded to deal with them through his illustrations. His solutions were precocious for the time and served to prove his analytical and perceptive skills. As Ferriss was commissioned by Corbett to produce his renderings of the zoning Ordinance, it is difficult to assess with any certainty where the influence of one individual ends and the other's begins. However, one can speculate that as Ferriss was the ultimate interpreter of the zoning laws he also had a great deal of power and control over both the visual and the technical content of his drawings. Corbett may have guided Ferriss through his calculations of the maximum volume or at what stage he was most likely required to limit the height of the tower in the drawings. But the visual effect of the drawings is only partially a resultant of this simple technical rigor. Speculating over these unknowns requires us first to gain a general familiarization with some of the significant elements which dictated the practice of design and the economics of skyscrapers. Confusion reigned in the early 1920s and it was only through the intuitive and visionary intervention of individuals such as Ferriss that both architects and urban planners could traverse the ambiguous and unclear first years following the introduction of the zoning laws and emerge triumphant.

In his drawings of the *Evolution* series, Ferriss shrewdly identifies the factors which determine the profitability of a tall structure. His findings are articulated in his writings. Identifying elevators as large consumers of valuable rentable space, Ferriss points out how unresolved set-backs are acutely damaging to the financial worth of a design. His text reiterates the findings of most experts: "The uppermost steps are of too small an area to be of use; when the spaces necessary for elevators and stairs have been set aside the remaining rentable area would not justify the expense of building."[25] An awareness of this fact dictates the transformations of the building mass as shown in his drawings. As the building progresses from its third to its fourth stage of development, the redundant steps in the setback blocks are eliminated to render a simplified massing in the final depiction (Figure 1.8). In order to refine his estimate of the factors that increase the economic value of a skyscraper, Ferriss takes his analysis one step further by considering the disadvantages of incorporating irregularities within the building volume. Irregularities, as they are caused by "the steps, because of their multitude and their comparatively small dimensions," complicate the process of assembling the structure, which according to Ferriss "would not prove an economical venture in steel."[26]

The economic productivity of a skyscraper was also a function of its height. Published in 1930 by an architect and an economist, J.L. Kingston and W.C. Clark, *The Skyscraper: A Study of its Economic Height* represents one of the most detailed analyses of the criteria that influence the economic value of tall buildings[27] (Figure 1.9). According to the authors, one universal rule applied to all skyscrapers: "No matter what the size or value or location of the plot or the character of the building, the law of diminishing returns will set in at some storey height and sooner or later a point will be reached beyond which it will not pay the owner to build under the existing conditions."[28] Although at times an exaggeration of the merits of the high-rise, this book convincingly portrays the skyscraper not only as the most profitable venture for developers and landowners of central sites, but also as the best solution for achieving healthy living conditions within the metropolis. In order to calculate the

1.9

Study of Economic Height: Eight Schemes in one diagram.

Source: W. C. Clark and J. L. Kingston, *The Skyscraper: A Study in the Economic Height of Modern Office Buildings.* Cleveland, New York: American Institute of Steel Construction Inc., 1930, p. 15

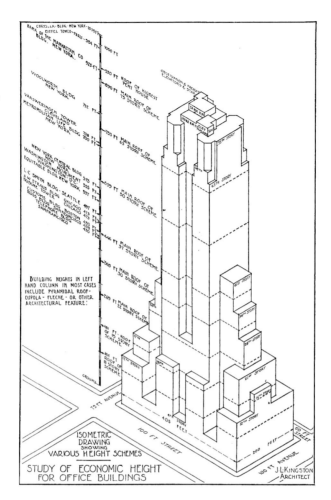

maximum economic height of a skyscraper on a relatively large site (200 × 400 feet), the authors determine the financial worth of eight separate schemes which are based on the same *parti*, lot size and land value and vary only in height.[29] Using the familiar technique of measuring the ratio of the gross cost to the net income, this study reveals that a sixty-three story building provides the highest rate of return on investments. The eight options are collapsed graphically into one mass and depicted as one drawing (Figure 1.9). Again, these drawings, though somewhat informative, do not provide the poetic need to capture the essence of the spaces created in the city based on these laws.

The visual proof

Ferriss's drawings presented their viewers with one of the first visual proofs of the economic advantage of investing in large sites. Incorporating a whole city block

Map or drawing?

within its boundaries, the large bulk of the building mass in these illustrations marks a precocious estimation of future conditions in New York City. Ferriss proved that not only were such interpretations realizable, but the expressive format in which his drawings were executed played a significant role in bringing the concept of big buildings one step closer to reality.

While his drawings provided architects with the conceptual groundwork to contemplate the design possibilities of large sites, Ferriss's visualizations posed a financial challenge to those venturesome businessmen of the early 1920s who believed in the risk factor of large commercial transactions which potentially multiplied profits.[30] Writing in 1929 about the economic hazards of small sites, Ferriss emphasizes that "if the site is quite small, it becomes impracticable to build any tower at all; on a very large plot, a tremendous vertical tower can, of course, be erected."[31] Clearly, a change in municipal zoning laws was something Ferriss sought – and something he ultimately got. In his article for *The New York Times*, Ferriss wrote:

> The most fascinating potentiality of the new laws may lie in their ability to finally bring forth the much-debated "new style" in architecture. Aside from those who seek a "new style" simply because they naively want something new, or naively want something national, there are designers who for sound reasons have found past styles inadequate to clothe present-day facts. . . . A new era commences.[32]

Also, in this article, Ferriss depicted his vision of New York in 1942, showing the effects of the zoning laws. This was one of Ferriss's first seminal pieces in *The New York Times*. Following this publication, a number of articles on Ferriss's drawings and visions appeared, most frequently in *The New York Times*. Through his associations with architects and the extensive publicity given to his works, Ferris transmitted his thoughts on the visual character of the architecture of the future.

The popularity of Ferriss's drawings

The popularity of his drawings became more apparent as more of his visions of the future city (Manhattan) were published in the popular press (Figure 1.10). In an article in *The New York Times* (1929), architectural critic R.L. Duffus stated, "In this vastly appealing book [*The Metropolis of Tomorrow*] he has assembled, in text and illustrations, his conceptions of what the builders have done and may do. . . . He gives us, first, the city of today, in order that we may see what trends are already at work."[33] Ferriss's popularity continued to grow through the success of his visions made apparent in his beautiful drawings.

In a pivotal article in *The New York Times* (1949), which described the future of New York's skyline in 1999, Hugh Ferris was one of the five regarded architects

1.10
Hugh Ferriss's drawing, showing his "Future Vision of New York's Business Centre." The author of this publication, R. L. Duffus, gives Ferriss great credibility, saying "Mr. Ferriss considers the problem created by the skyscraper" in the subtitle of the article. This was also around the time that Ferriss published his famed book, *The Metropolis of Tomorrow*, 1929. Duffus also cites in his article, "The Metropolis of the Future," that the drawing is taken from Hugh Ferriss's recent book.

Source: Ferriss, *The Metropolis of Tomorrow*, 1986, p. 113

of the time to be honored by being asked to depict their visions of the future.[34] In this article by Corbett, *The New York Times* commissioned five of the country's best architects to predict the future of New York; among these distinguished architects was Hugh Ferriss (the other four were Harvey W. Corbett, Robert Moses, Eliel Saarinen and Wallace Harrison). Also, in this article, Corbett dubbed Ferriss as a "Visual Consultant"[35] which indicates Ferriss's popularity at the time. Hugh Ferriss's vision braced the front headlines of this key article. His drawing of the future Manhattan was also used for advertising real estate predictions in the popular press:

> The artist has this advantage over the real estate man – with his imagination and his pen, he can picture in a few inches of space the New York of 2023; while real estate man must deal with the less romantic details of values, sales, mortgages, rentals, laws and locations. . . . It has taken only nine years to transform the business and residential heart of New York in even a more striking way than Hugh Ferriss has pictured it above.[36]

Map or drawing?

Ferriss's zoning Ordinance drawings received broad exposure.[37] They were showcased in February 1922 at the annual exhibition of the Architectural League of New York and published in its catalogue, and were "singled out for praise by a number of critics."[38]

As an acute observer of dominant architectural movements and a trusted member of architecture and urban design committees, Ferriss participated in current debates and played a significant role in shaping the post-war architecture of American cities. The dominance of sculptural forms and the suppression of historicist styles as they appear in Ferriss's visions coincide with the two characterizations which architectural historians use to describe the most prominent traits within the post-zoning architecture of New York City.[39] That Ferriss was able to isolate these elements at such an early date gives his visions credibility. Ferris showed how architects could work constructively within the tight parameters established by city ordinances and zoning laws. He initiated a new chapter in American architecture because of his time spent as a noted lecturer at other leading universities and because his writings and drawings changed the ways in which people looked at cities.[40] He was and remains famously known as "the delineator of Gotham."[41]

"The new style blended with the brooding visions of Hugh Ferriss, an architectural visionary whose influential drawings conjured a futuristic Gotham of mountainous skyscrapers connected by elevated superhighways," in the words of architectural critic Steven Litt.[42] Ferriss was not only seen as a "consultant," [43] but a "portrayer of the city of the future."[44] He was also elected to head the New York Chapter of the American Institute of Architects.

The influence of Ferriss's map-drawings

Ferriss influenced the architects of his time in less apparent ways. As implied earlier, the requirements of zoning together with the financially driven demands of clients had necessitated new approaches to design. The drawings of Ferriss gave visual proof of alternative ways through which architects could tackle design criteria which, following the enactment of zoning laws, had become immensely complicated. In his illustrations Ferriss had demonstrated a process for designing a tall building that was particularly suitable under the new circumstances. His was a reductive method in which he sculpted the maximum bulk of the structure to arrive at the most practical and aesthetically pleasing solution. The method was particularly useful due to its flexibility which permitted all the requirements of zoning and cost factors to affect the form, while still leaving room for the architect to intervene. While the lack of textual evidence does not permit an absolute statement, it is likely that many architects took inspirations from the artist's methods of delineation and imitated his process in various media. A plasticine model made by New York architect Walter Kilham demonstrates the maximum envelope of the Daily News building (1927), and bears a very close resemblance to the first illustration made by Ferriss to illustrate the maximum building envelope in his imaginary skyscraper. Through a subtractive process, Kilham carves the form using a knife. In essence, this model became a sculptural map in which Kilham was able to see the three-dimensional parameters of the law.

Amongst those familiar with Ferriss's illustrations, architects can be identified as the group of professionals who arguably absorbed the greatest influence from these works. The speed and ease with which his drawings were assimilated within the architectural discourse of the period is evidenced by considering two facts about the zoning Ordinance and its reception. First, the new laws on zoning had been primarily based on practical concerns with no consideration given to their aesthetic dimension; and second, the legal text of the Ordinance was recognized by most of its users as lengthy and confusing.[45] Addressing these concerns, Ferriss's renderings not only helped to clarify the abounding confusions about the laws but also set the grounds for contemplating their aesthetic and spatial possibilities. The actual buildings and Ferriss's drawings look similar because the forms were necessary consequences of the Ordinance in both cases. The published works of Ferriss which included the 1922 drawings of the *Evolution* series, as well as many other pieces used either as parts of advertisements or in exhibitions, attracted a wide audience and became instrumental in pointing architects towards new and exciting directions.

The aesthetic potentials of the Ordinance were recognized by American architects of the period, and can be seen addressed in architectural publications. However, the majority of the articles in which the relationship between zoning and aesthetics are discussed only begin to appear in the early 1920s, a few years after the enactment of zoning laws in 1916. As Carol Willis notes, this date coincides with the mid-1920s boom in commercial construction, during which period architects also began to explore the formal possibilities of zoning in actual projects.[46]

Contrary to what might be expected as a consequence of laws that place restrictions on the professional and creative practice of designers, the response of the architects to the zoning Ordinance as represented in the publications of the period is marked by a high level of enthusiasm and excitement.[47] The introduction of the zoning laws intrigued architects with their potentials for a new architectural style. Not only praised for their capacity to improve the quality of life within the city, the laws' strictures were viewed as an opportunity to re-evaluate the state of architecture. They would help architects in their quest to identify an appropriate style for their time. As they only stressed practical concerns, the new laws provided architects with the ammunition to challenge the dominating norms, which favored the adoption of old styles and motifs, while guiding them in their mission to create an altogether "modern" brand of architecture that was in tune with the architectural movements of the period.[48] The new style of architecture would abandon the primacy of the façade and compositions based on "cubes," and instead allowed an approach to design methods that took advantage of the three-dimensionality of the pyramid.

Architects were not alone in the early 1920s in welcoming zoning for its potential to enhance urban life and aesthetics. The high surge of reaction towards the Ordinance and its implications for architecture and urban design found its voice in numerous articles published in the widely circulated popular press, such as the *New York Times* magazine. The frequency with which articles on zoning appear through the 1920s signals the attractiveness of this topic amongst the public. Illustrations by Ferriss frequently accompany these publications to give visual and graphic credence to the often extolling text in which zoning is celebrated for all its positive intentions.[49] That journalists used illustrations by Ferriss to give support to their arguments attests to the visual power of the artist's drawings which so effectively captured their audience. It also substantiates the role of his drawings in the communication of intangible and yet important facts about the city. The optimistic attitudes of the general public towards the architectural effect of the zoning Ordinance in its beginning stages are highlighted in a passage from a 1924 article in *The New York Times*:

> It might not be a far-distant tomorrow when cubical chaos is dethroned, the reign of Harmony and Symmetry inaugurated. Then, instead of streets hemmed in by boxes of varying heights, rows of structures, each row of an uniform height at the street line: instead of a skyline of clashing cubes, a skyline of colossal steps, the upper level of every street and avenue a vista of diminishing terraces.[50]

Words such as "harmony" and "symmetry" which appear in the above quotation are frequent in architectural writings of the early 1920s, and give a sense of the qualities which architects of the period strove to create in their works. Often using contradistinctions such as "cubical chaos" against "stable," "three-dimensional" and "pyramidal forms," the texts highlight the differences between the status of architecture in its pre-zoning days and its new character.[51]

Still others supported the Ordinance for its promise to elevate the position of architects, and by implications the arts, helping to redefine the professional hierarchies within the building industry which prior to the enactment of the laws was dominated by developers. In a mid-1920s article that discusses the influence of Ferriss's drawings in relation to the architectural aftermath of zoning, the author identifies the laws of the Ordinance as the means by which architecture is to be emancipated from the "smart commercial men [who] were quick to realize that a house without 'profile' needs no architect to design it."[52] He elaborates:

> But now of a sudden, certain purely practical measures, namely, the new zoning laws in New York, have brought the architect not only back into his own, but into a prominence he had not occupied since the days of the Renaissance in his association with the corporation of activities which make the trade of building. . . . The strange part of the business is that these laws, which had no concern whatever with aesthetic considerations, have had a tremendous aesthetic reaction, and for their

complete and successful application the cooperation of the artist-architect is now more necessary than that of the engineer-architect.[53]

The three quotations above accentuate and give proof of the positive vibrations which the zoning acts of 1916 inspired in architects and the general public, beginning from the early to mid-1920s – prior to an awareness of the threatening dimensions of the new laws. What is important to emphasize is that while architects and designers of the period were quick to recognize the aesthetic and spatial possibilities of zoning, no one had a clear image of the precise consequences of these laws as they affected architectural forms and urban conditions. While people's understanding of zoning was mostly based on their incomplete apprehension of its requirements, their collective enthusiasm was supported through a shared sense of excitement which masked, or plainly ignored, the negative ramifications of its laws. Exploring the three-dimensional reality of zoning through drawing and building remained as tasks for future architects and artists.

As one of the earliest explorations of the aesthetic consequences of zoning, the drawings of the set-back building envision architecture in the future. In these works, the characteristics which define architecture for a "new era" are prophesied and communicated as visual information. Through his interpretations of the zoning laws, Ferriss constructs a philosophy of design according to which the clear articulation of a building's sculptural qualities best expresses its architectural character. He refines his definition by advising against the insistence of some architects to employ old styles as sources of inspiration. Ferriss's stance can be heard loud and clear in his writings.

The commutative illustrations

Ferriss's illustrations helped to clarify the content of the zoning Ordinance, while leading their viewers to consider both aesthetic and urban design consequences of the legal codes. Ford had, prior to Ferriss, produced some of the earliest visual representations of the effects of zoning (Figures 1.2 and 1.5). Seemingly appropriate for explaining data in a direct and simplified manner, Ford's linear and crisp style, however, was scarcely enough to allow his illustrations to communicate their message.[54] The factors which enabled Ferriss to attain greater success with his drawings than Ford does with his are understood by considering the artistic input in each case. Equally accurate and as truthful to the taxonomic structure of the zoning Ordinance as Ford's diagrams, the drawings of Ferriss are also superb works of art.

In his illustrations of the zoning codes, Ferriss applied the genius and technical expertise of an artist to the banal legal content of the Ordinance in order to highlight significant instances within the laws as they shaped and defined the built environment. Highly dramatic and forceful, his images captured the imagination of their audience with their graphic and artistic merits. The drawings' great influence and the graphic potency with which they affected their audience can be abstracted and seen as the product of two significant attributes. Introduced earlier in this chapter, these two qualities can be recalled by the adjectives "seductive" and "revelatory." These characteristics can be distinguished from two other definitions, remembered as "informative" and "suggestive," by considering the ways in which each of these attributes functioned in Ferriss's illustrations to energize and explain their subject. While the latter two categories were instrumental in negotiating the legal content of the Ordinance as visually intelligible documents, the former increased the drawings' potential and impact to inform and communicate their message. To further clarify, we may consider that while the informative and suggestive attributes primarily relate to the audience's intellectual faculties, the seductive and revelatory attributes appeal to their emotions.[55]

Drawing methods and media

In an analysis of the ways in which Ferriss's illustrations of the zoning laws functioned as both informative and visually captivating artifacts, a few questions may be asked. Precisely what were the methods by which the artist translated the legal codes into highly effective visual documents? What were the techniques with which he infused his drawings with such graphic force? And lastly, to what extent can the drawings be understood as a direct outcome of the legal text, and to what extent were they a product of the artist's own personal interpretations of the codes? Valuable clues about the techniques and approach of the artist to his drawings can be found in his writings on the topic of rendering, which primarily focus on the tasks and responsibilities of an architectural delineator – a position which Ferriss invested a great amount of his time and energy to define. The illustrations of the Ordinance also constitute a significant source of materials which can be examined to reveal insights into the artist's methods. Particularly important to investigate are first, the artist's use of the drawing medium; and second, the degree to which he was willing to manipulate his subject matter – whether it was the legal codes of the Ordinance or just a typical rendition of a building – in order to draw out its most essential components.[56] His famous Ordinance drawings were published in the 1929 *Encyclopedia Britannica*.

The heavy chiaroscuro effect of charcoal is specifically suitable for emphasizing the massing of a building, while masking its details. The effort to highlight the sculptural features and masses within a building runs parallel to the architectural ideologies which Ferriss, along with some of the progressive architects of his time, was promoting as the future architecture of America.[57] In his entry on "Rendering" written for *Encyclopedia Britannica*, Ferriss draws attention to the expression of "mass" as one of the first and most essential tasks of a renderer. He begins: "From the renderer's point of view a building is, in the first place, a material mass . . . the renderer must realize the presence of mass before he can fully realize the presence of any appurtenant form."[58] To reinforce a point made earlier in this chapter, just as isolating the masses in a building gives it architectural credibility, "the first necessary attribute of a convincing architectural rendering is, correspondingly, an adequate suggestion of mass. Without this primary effect of solidity, all details which may be delineated later must appear without body and the presentation as a whole must lack substance."[59] In another article he imagines a renderer who is about to make his first strokes: "the draftsman standing before the easel, made the assumption that, for the moment, the paper represented *space*. With the intention of introducing into this space, the presence of *mass*, a number of lines were lightly sketched."[60] Process drawings on how Hugh Ferriss draws are captured brilliantly in his documentation of "progressive views of a Rendering" published in *Encyclopedia Britannica*, 1929.[61] This figure depicts the initial drawing setup (with guide lines) tentatively suggesting a mass, followed by additional lines added as material for tone and value. The subsequent steps show more lines, tones and shade added, using kneaded eraser and paper stump.

Throughout his career as an architectural delineator, Ferriss developed a very idiosyncratic style of drawing. Using charcoal as his primary medium of expression, he emphasizes in his renditions the sculptural qualities of buildings. The choice of charcoal is consistent with the architectural effects he intends to create, while the medium enables him to develop a particular method of drawing in which he approaches his subject matter with great care, only bringing out the essentials through a gradual and sequential process of adding and subtracting using a kneaded eraser. Charcoal supplied Ferriss with an appropriate medium through which the effects of zoning were dramatized and rendered as seductive and atmospheric images to attract and influence the thinking of their viewers.

The suitability of the drawing medium for both creating a powerful visual effect and communicating an architectural vision establishes a particularly strong link between the message of a drawing and its technique of production. The sculptural juxtaposition of the masses in a building embodies its essential character and the application of charcoal facilitates the negotiation between a thought and its expression as a drawing. The monochrome yet dramatic visual effects, especially suitable to charcoal, also benefit the illustrations of the zoning laws (Figures 1.1, 1.6, 1.7 and 1.8). Not only do these conditions metaphorically invoke honest and unpretentious sentiments which fit well with the practical demands of zoning, they also substantiate Ferriss's interpretations as credible and accessible documents. The dark tones of the drawings, together with the manipulation of the light source, invoke a seductive yet eerie portrait of the city.

The artist's graphic techniques for producing a desired effect are not limited to his masterful use of the medium. They are also a result of the way in which he approaches his subject matter. Reviewing the artist's writings on the art of rendering reveals that although Ferriss was hard pressed to capture the reality of his subjects, he was also willing to bend the rules and conventions of drawing, and sometimes even to distort the geometry of buildings and their context, in order to bring out their most essential components. Throughout his writings, Ferriss emphasizes the importance of capturing the "essence" of a building. He identifies "one of the chief concerns of the renderer to comprehend the nature of the architectural idea which his subject embodies, the trend of thought the architect has expressed."[62] An architectural rendering that captures the "essence" of a building is termed by Ferriss as "truthful." Ferriss's definition of a truthful rendering, however, surpasses a representation that merely depicts the physical traits of a building.

While not one to promote "exaggerat[ing] certain factors of a proposed building in order to create a more favorable impression . . . or to advertise," Ferriss is often willing to sacrifice some aspects of reality if it means gaining in "truth."[63] As it relates to the articulation of building components, he believes that some elements within a building merit special attention and therefore need to be expressively pronounced, while others should be subdued. Expressions of materials are for example important to a truthful rendering, while details and the treatment of fenestration are not. Consequently, Ferriss tolerates a great deal of artistic license in order to bring out the essence of a building. Discussing the extent to which a renderer is permitted to eschew facts in order to depict the building as "truthfully" and "completely" as possible in a single drawing, Ferriss assumes a hypothetical situation in which a renderer has been asked to delineate a building from a viewpoint which is inappropriate for revealing a very significant aspect of the structure. His response only accentuates how far the artist himself was willing to stretch the limits of objectivity: "In such a case, it would appear that he [the renderer] is not so much permitted as actually required to slight incidental facts of his viewpoint in favor of the essential facts of the subject which he is viewing."[64] Ultimately, an ideal rendering is, according to Ferriss, one that conveys "the material, the emotional and the intellectual facts in the same rendering."[65] Ferriss's *Evolution* drawings spoke a truth about the law and exposing a new city at hand.

While a universal set of criteria can measure the success of an architectural illustration (such as emphasis on massing), the intuitive process of a drawing exercise as it is directed by the artist/renderer enhances his productions with subjective and personalized dimensions. Classifying the different factors which lead to an accomplished rendering, Ferriss emphasizes the presence of the artist as the ultimate creator of the finished product. A passage from an early article entitled *Truth in Architectural Rendering* defines the role of the renderer as both a spontaneous and responsible arbiter who is required to make sound judgments:

> Buildings possess an individual existence, varying – now dynamic, now serene – but vital, as is all else in the universe. If we may not, here and now, transcribe this existence in our paintings and drawings, we at least may – we must – choose *some aspect* of the entire truth as our subject. The importance of our work will depend on how we choose.[66]

The freedom of the artist to choose some aspect of the entire truth as its embodiment also authorizes the artist with the right to question and redefine the very mechanisms and conventions of rendering. Bending the conventions of architectural drawings is acceptable to Ferriss as long as it does not threaten the status of a drawing as a reliable document. On a number of occasions, Ferriss challenges absolute adherence to the canons of architectural drawings as they were practiced by most contemporary draftsmen. He questions the validity of impressions that are entirely based on a predetermined set of rules. The conventions of making a perspective drawing are a common target: "It is generally taken for granted that if accurate floor plans and elevations are available, an accurate image of the building can be produced by following the rules of perspective draftsmanship."[67] Ferriss objects to an uncritical reliance on the rules of perspective solely based on the conception that they are scientific. While perspective is recognized as useful because of its accuracy, it is only a convention and can be a hindrance to the creative process.[68]

It is quite evident that Ferriss was not satisfied with a literal translation of his subject matter. He sought to express what was most essential in his subjects. A mathematically accurate representation did not necessarily convey the perpetually changing character of objects as they appeared in real life – impressions which Ferriss was intent on capturing in his drawings. However, he did stress some measure of accuracy. Only an accurate image fulfills the first three objectives of a rendering, which according to Ferriss must "convey advance realizations of proposed structures, aid in crystallizing ideas in the architect's mind and interpret the architectural significance of existing structures."[69] A rendering first and foremost had to be a reliable visual source to be scrutinized by architects, engineers or clients.[70] However, as the communication of facts constitutes the first responsibility of a draftsman, only an experienced and talented artist is to be trusted with the task of selecting what is to be omitted and what calls for emphasis in a rendering.

Ferriss does not discuss in detail his methods of depicting the Ordinance of 1916, but one can see the technical mastery of the artist which enabled him to balance legal information with his own visionary and fictional input to create precise and highly provocative documents. Using the content of the Ordinance as reference, the artist proceeded to consider the possibilities of using the drawing medium and techniques of representation, which sometimes involved manipulating his subject matter in order to effectively communicate the thrust of the new laws. In Ferriss's work the medium and technique of drawing are intimately in unison with the effects that he is seeking to create. A few observations made about the drawings can reveal the ways in which the artist applied the techniques of drawing to convey the message of the codes in an emphatic manner.

Focusing on the drawings, the location of the vantage point is critical. It allows a full view of the three-dimensional translation of the legal codes. The rules of perspective allow the artist to mark a significant point within the transformation of the building mass as it reflects the effects of zoning on the economic height of the tower. In the first two drawings, in which the tower rises indefinitely as allowed by the zoning specifications, the shaft of the tower shows no effect of vertical perspective and its lines are parallel, emphasizing an infinitely high tower (Figures 1.1 and 1.6). In the last two drawings, the vertical lines of the masses and planes of the tower are slightly tapered to suggest a definite height (Figures 1.7 and 1.8). Nevertheless, the receding lines express the soaring height of the skyscraper.

In order to record real-life impressions and to show major transformations within the building mass, Ferriss uses light to create a scenic effect. The medium of charcoal is particularly appropriate for emphasizing light conditions within the drawings and giving depth to the space. The masterful use of lighting also brings out the masses. In the first two drawings where building mass is still at the beginning of its evolution, reflecting only the legal requirements of maximum volume, the source of light remains unclear. The building in the first illustration also lacks context and the mass of the building as a direct outcome of the zoning requirements stands monumentally in a *tabula rasa*. As the building volume is further articulated, primarily as a function of its environment, a context is also indicated and other structures begin to surround the building. The definition of context evaporates again in the last drawing, both as a consequence of the shifting position of the light source and the intention of the artist to isolate the final resolution as the most significant object within the picture.

At certain places within the images, the masses merge as a result of the equal tone that is applied to the planes at different depths. This condition is not so pronounced in the zoning drawings but can be seen in other works. The effect created by this condition is a lack of clarity; the viewer is not so sure where the mass ends and space begins. At such points, the viewer is only offered a suggestion of the forms and left to imagine the resolution of the masses. In "Building in the Modeling," the handling of the medium and visual elements describe transparent, crystal-like qualities of the forms (Figure 1.11). In this drawing, the stepped masses of the building suggest flexible urban conditions. The background overlaps the foreground as the lines of the building in the distance pass through the lozenge-shaped outline of the roof of the front building. While the building mass within the figure in the distance is clearly defined through its perimeter contour lines, towards the top the lines are eliminated and the surfaces interconnect. The drawings are left intentionally unfinished. They allude to facts and yet leave room for personal interpretation.

Every time Ferriss composes an image, it is a distinct conception. While the zoning guidelines provide the backbone to his illustrations and the uniting thread through which they are strung together, every image is a separate and isolated piece of art. The effects of light enable each image to gain an autonomous presence. As segments of the surface are pulled out and highlighted through the shifting position of light, each interpretation of the city becomes a new vantage point that reveals yet another dimension of the future city. As a result, each drawing, meticulously rendered, only shows what is most significant in the mind of the artist about the zoning laws and the architectural features of the building. The choice and handling of the medium only add to this effect, while the subject matter is accordingly

1.11
Ferriss's sketch "Building in the Modeling." The stepped masses of the building suggest flexible urban conditions.

Source: Ferriss, *The Metropolis of Tomorrow*, p. 85

adjusted. For instance, when Ferriss has in mind the direct relationship between form and the laws, the drawings describe the buildings as isolated objects; when the artist is concerned about the effect of the laws at an urban scale, his drawings take on a much more abstract character in which the masses placed together describe a landscape that is completely foreign to the conventional image of a city. The potentials of the zoning laws are only amplified through a conglomeration of mountain-like masses with their edged-shaped building contours as exemplified in one drawing titled, "Crude Clay for Architects" (Figure 1.12). In this image, the building material is unclear and the application of charcoal again heightens the sense of mystery through the juxtaposition of dark masses in the foreground with the middle section illuminated. The lighted figures give the masses a truly ephemeral quality that recalls that section of Ferriss's description in which the artist recognizes the molded forms as only an unfinished representation of the building that is in need of articulation at the hands of the architect.

1.12
Ferriss's sketch "Crude Clay for Architects."

Source: Ferriss, *The Metropolis of Tomorrow*, 1986, p. 83

Exposing the city

The *Evolution of the Set-back Building* drawings are best described as portraits of a new city. They definitely exposed the new conditions and urban form of New York City that is appreciated today. They are portraits because in them one can read not only the visual translation of legal constraints, but a grand vision of a future architecture. Had the artist only depicted factual elements, his drawings would not fulfill their role as seductive and revelatory artifacts. With sections left unfinished, the images intrigue the imagination of their audience with further possibilities and architectural forms. The drawings inspire and draw the viewer towards them as their subject matter invokes a theatrical mood that is both accurate and elusive. A dialogue is initiated between the drawing and the viewer. In order to energize this dialogue, both the content and style of representation play key roles.

New York architect and critic James Sanders comments on the influential impact of Ferriss's zoning drawings:

> The impact of architectural rendering grew even more pronounced in the modern era. In 1916, New York passed an innovative zoning law that sought to preserve a measure of light and air in the crowded city by requiring that buildings step back, like a ziggurat, as they rose into the sky. But it took a series of charcoal drawings by the brilliant illustrator Hugh Ferriss, published in The New York Times in 1922, to show the thrilling architectural consequences of the law. Indeed, as the new breed of towers began to rise, Ferriss's moody drawings of their mountain-like masses, terraced setbacks and soaring pinnacles proved as crucial as any built work to setting the tone for New York's super-charged urbanism of the 1920's.[71]

The enactment of the zoning Ordinance in 1916 promised a drastic reconfiguration of the built environment in New York City. While the set-back rules were deemed to improve living conditions for the urban dweller, the limitations on the bulk and height of buildings both complicated the practice of design and threatened the economic gains of landowners and developers. By translating the legal parameters of zoning into visual documents, Ferriss contributed to an understanding of the zoning laws. As "maps," his drawings functioned as media through which abstract data bearing directly on the spatial definition of urban conditions were explored and visually displayed. In re-evaluating the term "map," Ferriss's drawings are definitely maps, ones that guide the architect and city dweller into a new city based on the law. The first drawing of the *Evolution* series was a delineation of the maximum envelope of the building as specified by the zoning laws. As the starting point within the process of designing a skyscraper, this drawing revealed the immediate impacts of zoning on the practice of architecture. It meant a whole new approach to design. The next three drawings reveal the formal consequences of zoning as they reflect economic and aesthetic considerations. No longer permitted to maximize the rentable floor space in their investments, landowners and speculative developers benefited from Ferriss's drawings as sources through which they could visualize the restrictions of zoning. They were allowed to maximize rentable floor space within the provisions of the Ordinance.

As one of the earliest explorations of the aesthetics of zoning, the *Evolution of the Set-back Building* drawings enabled their audience to consider the possibility of interpreting the zoning laws in order to create a new style of architecture that was in tune with the period. Surfacing at a critical moment within the development of post-war architecture in America, Ferriss's drawings were instrumental in giving a formal definition to the thoughts and ideas of those architects whose concerns with the quality of life in cities inspired them to celebrate the introduction of zoning as the practical measures they were seeking. The best proof of the influence which Ferriss exerted through his drawings can be found in the architecture of the period. A dominance of unadorned masses and a defiance shown towards adopting historicist styles characterize prominent architectural productions from the mid- to late 1920s in New York. Both these attributes are found in Ferriss's 1922 drawings and establish a firm connection between his imaginings and the preferred architectural style of the decade.

The power with which Ferriss affected and captured the imagination of his audience is only partially explained by considering the communicative and visionary dimensions of his works. His drawings were also artistic masterpieces. In his illustration of the Ordinance the rigor with which the legal elements of zoning were pursued is only matched by the vigor and expertise that the artist applied in managing his drawing medium and methods of execution. The attraction of his illustrations only strengthened and facilitated the reception of their message.

While Ferriss was in his drawings primarily concerned with the technical substance of zoning and its economic and architectural consequences, the artistic dimensions of his renderings expanded both their relevance and intelligibility beyond the reach of experts to include the general public. His medium (charcoal), his technique of shade, shadow and light, and the sheer size of the drawings inspired reactions from public officials, architects and urban dwellers. Even the ordinary person was able to access and respond to his works. The seductive character of his drawings was responsible for increasing their commercial value and prepared them for public consumption. That the drawings reached a wide audience is attested to by the wide publicity that Ferriss received through the popular press such as *The New York Times* magazine, the numerous exhibitions of his works,[72] (including Ferriss and Corbett's 1925 Titan City Exhibition in New York in October of that year and the Anderson Galleries exhibition in February 1925) and the use of his drawings by companies such as the American Institute of Steel Construction to advertise their products. In his obituary published in *The New York Times*, Ferriss was described as "Farseeing designer . . . an architectural visualizer of amazing imagination and *influence* . . . his work still is ahead of his times."[73]

The power of Ferriss's famous "four-stage" drawings are supported in Willis's article:

> In these images Ferriss gave the setback formula an almost iconic identity. Widely published and exhibited throughout the 1920s, these drawings were of unparalleled importance in impressing contemporaries with the power, beauty, and inchoate modernity of the unornamented setback form. . . . Apparently inspired by Ferriss's grandiloquent visions, Corbett soon began to echo the delineator's optimism.[74]

The importance of Ferriss's zoning Ordinance drawings are further emphasized in Willis's article:

The dramatic renderings of Hugh Ferriss, today the best known of this body of futuristic speculation, were extremely influential in shaping the thinking of his colleagues. . . . [In] 1929 he published a collection of his verbal and visual prophecies in his masterwork, The Metropolis of Tomorrow. . . . Both the forms of Ferriss's colossal setbacks and their disposition in a comprehensive city plan owed their inspiration to the New York zoning law.[75]

The early twentieth-century architectural critic Claude Bragdon writes about Ferriss's influential status in a 1931 article in *The American Mercury* entitled "Skyscrapers":

The final influence to be noted is that of Hugh Ferriss, who is not, in the strict sense, an architect at all, but an architectural "renderer." His structural sense and his aesthetic sensitiveness are however so much greater than those of many architects who employ him, that it is often highly disillusioning to compare his drawings with their originals.[76]

Ferriss's influential status is further re-enforced in Bragdon's article:

Ferriss's *The Metropolis of Tomorrow* appears to be exerting an influence on skyscraper design in a manner analogous to that which Piranesi's prison etching exercised upon the design of George Dance's Old Newgate.[77]

Bragdon also expresses how architects like Le Corbusier and Ferriss have advised that "skyscrapers should be built only at focal points where traffic lanes converge, and the surrounding buildings should be kept relatively low, or else broad avenues flanked by high buildings should alternate with streets flanked by low ones. In this way light and air for all would be assured."[78]

Through both his visual and textual impressions of the metropolis, Ferriss clarifies his stance towards the much debated public dilemma over the position of man whose existence was rendered inconsequential within the city of towers. The sensitivity with which Ferriss deals with this subject unites him with the ideological standpoint of the early reformers, responsible for thinking up zoning as a way to combat urban congestion and unhealthy living conditions, and a handful of architects whose concerns about the presence and value of human life inspired them to support zoning in its early stages. Architecture was for Ferriss a means through which man related to his built environment. He believed: "If there be anything in the theory that the building affects the man in the street, we may indeed regard this architectural development as possessing human interest."[79] Therefore, architects were laden with a much heavier responsibility than was previously thought of as their job. Architecture impacted on the very perception of reality. Designers were required to be fully cognizant of the psychological and emotional dimensions of their creations. As Ferriss explained:

> It has been our habit to assume that a building is a complete success if it provides for the utility, convenience and health of its occupants and, in addition, presents a pleasing exterior. But this frame of mind fails to appreciate that architectural forms necessarily have other values than the utilitarian or even others than those which we vaguely call the aesthetic. Without any doubt, these same forms quite specifically influence both the emotional and the mental life of the onlooker.[80]

While Ferriss does express his reservations about such an idealistic reality, in conclusion he asks: "Are we to imagine that this city is populated by human beings who value emotion and mind equally with the senses, and have therefore disposed their art, science and business centers in such a way that all three would participate equally in the government of the city?"[81]

Ferriss's drawings created compelling impressions of the metropolis. Whether his work was seen as haunting, dramatic or provocative, one can speculate that Ferriss's credibility, along with his seductive portraits of the city affected the perception of zoning and its consequences within the city. They expressed a prophecy of the city that caused both architects and public officials to contemplate the effects of the laws. Ferriss could only have produced this great effect with the emotional fervor of an artist, which also enabled him to capture the spirit of his time. This is re-enforced by Paul Goldberger:

> Hugh Ferriss was our century's most potent architectural renderer, a man whose richly shaded drawings evoke a startling degree of romance, power, drama and hope. Ferriss had, in a sense, a double career. He drew the skyscraper designs of other architects with such eloquence that his renderings have become the standard visual representation of many buildings. But more important than Ferriss's work in the service of other architects' schemes was his use of the pen to reveal his own visions of what the city might be. More than any other American artist or planner of his time, Hugh Ferriss foresaw the great impact the skyscraper would have on the American city, and sought to give the high-rise city coherent and civilized form. . . . [His drawing] was organized, but not regimented; it had both softness and strength.[82]

Ferriss questioned all conventions, including the ones which structured the profession of renderers: When is something a work of art and when is it merely statistics? This is a topic of investigation throughout the thesis: what visual means are necessary to convey both artistic and informative properties as they relate to mapping? Ferriss was able to suspend his works within the two realms, and it was precisely the tension – the tension between the expressed and the unexpressed, between fact and fiction, between idealisms of an ordered and perfect future and practicalities of design – which fueled his works with great visual and communicative force.

Notes

1. Maps are graphical representations that facilitate a special understanding of things, conditions, processes, or events in the human world. (J. B. Harley and David Woodward, *The History of Cartography*. Chicago: The University of Chicago Press, 1987.) Maps are visual representations of the world or a part of it. They are visual instruments that guide the individual through the city or other part of the world they depict. Usually two-dimensional, the conventional road map depicts elements such as streets, landmarks, green spaces and water features on a paper plane. In 1911, *The Century Dictionary* defined the term map as "n. a drawing upon a plane surface representing a part or the whole of the earth's surface or the heavens, every point of the drawing corresponding to some geographical or celestial position, according to some law, of perspective, etc., which is called the projection, or, better, the map-projection." (William Dwight Whitney, *The Century Dictionary*. New York: Century, 1911). In 1927, the term map is defined as "n., A representation, usually on a flat surface, of a part or the whole of the earth's surface, the heavens, or a heavenly body; hence, a map-like representation of anything." (ed. H. G. Emery and K. G. Brewston, *The Century Dictionary of the English Language*. New York: Century, 1927).
2. The comparison between the works of Ferriss and those of Piranesi and Boullée is a popular theme in discussions of his work. See, for example, Jean Ferriss Leich, *Architectural Visions: the Drawings of Hugh Ferriss*, p. 17.
3. Benjamin Forgey, "The Poet of the Skyline: Hugh Ferriss and His Projections from the 1920s" in *The Washington Post*, 28 February 1987: G1.
4. Christopher Hume, "Unfilled Promises of the '20s Hits Home at Metropolis Exhibit" in *The Toronto Star*, 22 June 1991: F1.
5. Laws on zoning affected the interests of many organizations and individuals within the city, and legal recognition of the merits of zoning only materialized at a time when the prevailing conditions stressed the commonalities, and *not* the conflicting positions of such groups as represented by property owners and city planners (the Supreme Court had found building height regulations constitutional in 1909). While issues of public health and enhanced living conditions for the urban dweller can be identified as the true motives behind the actions of committees and activists who first instigated the debates on zoning, property owners and businessmen whose efforts were most effective in passing the laws on zoning only used such ideological reasoning as a guise to hide their economic interests. In its later stages of evolution, zoning became the domain of speculative developers, business executives and property owners. For details of the politics of early zoning efforts see, "Regulating the Landscape: Real-Estate Values, City Planning, and the 1916 Zoning Ordinance" in David Ward and Olivier Zunz, eds., *The Landscape of Modernity*, pp. 19–42; and "The 1913 Report" in Seymour I. Toll, *Zoned American*, pp. 143–171.
6. Carol Willis, *Form Follows Finance: Skyscrapers and Skylines in New York and Chicago*, pp. 41–42.
7. One important catalyst for the Ordinance as it is discussed by the majority of architectural historians was the building of the Equitable Life Insurance Building (1915) at 120 Broadway in the heart of the lower Manhattan financial district. This building, which covered a whole block and rose to a height of 39 stories, was a massive undertaking with 1.2 million square feet of rentable space that made it the largest office building in the world. Casting long

shadows on surrounding structures, this building raised many complaints from neighboring landowners whose land values had been tremendously affected by the new construction. Built primarily as a money-making machine, the Equitable building produced a 5 percent return on the owner's investment. This was a high figure as for example when compared to the Woolworth building which only earned a 2.5 percent return. Paul Goldberger's characterization of the Equitable building as "nothing more than a device to cram more floors into the sky . . . built not as a monument, but as economic adventurism," echoes the impression of most critics. Paul Goldberger, *The Skyscraper*, p. 15. This was also reinforced in Hugh Ferriss's article, "The New Architecture" in *The New York Times*, 19 March 1922: 54.

8. McAneny, Manhattan borough president, is the author of this text. Heights of Building Commission, *Report of the Heights of Buildings Commission to the Committee on the Height, Size and Arrangement of Buildings of the Board of Estimate and Apportionment of the City of New York.* New York: City of New York, Board of Estimate and Apportionment, 1913, p. 1.

9. The Ordinance divided the city into three use districts, residential, business and unrestricted, and five height districts. The most demanding restrictions applied to areas in Manhattan's central business district (CBD), as well as business sections in Brooklyn, and waterfront areas in Queens and the Bronx. The last requirements – area restrictions – prevented new buildings from covering their entire sites. In the CBD, these restrictions were again most demanding.

10. The sections of the Ordinance in which height restrictions were discussed attracted the greatest amount of attention particularly from architects. As architectural historian Carol Willis notes: "although the idea of districting had been the principal interest of the professional planners who led the zoning movement, it was this limitation on the form of tall buildings that particularly stimulated the thinking of many architects in the 1920s." Carol Willis, "Zoning and Zeitgeist: The Skyscraper City in the 1920s" in *Journal of the Society of Architectural Historians*, March 1986, pp. 47–59. Herbert S. Swan, Executive Secretary of the Zoning committee, explains this phenomenon based on these sections having the most visible consequences. "It is quite natural that the provision in the New York Ordinance to attract the most widespread popular attention should be that regulating the height of buildings. As the time since the adoption of the ordinance lengthens, and the number of skyscrapers erected under the law multiplies, the more conspicuous is the effect of this provision. Each new high building, because its façade must be set back in steps, terraces or mansards with each unit of increased height, tends only to enhance the public's interest in the height regulations of the ordinance." Herbert S. Swan, "Making the New York Zoning Ordinance Better: A Program of Improvement" in *Architectural Forum*, October 1921, p. 128. It is important to note this since it shows that Ferriss formulated his drawings directly in response to the prevalent demands of his time.

11. Each district had a different height restriction expressed as a ratio between the typical width of streets and the maximum permitted height. For example a 1:1½ district which applied mostly to less dense areas would imply that the building was allowed to reach a height of 150 feet if the street fronting it was 100 feet wide. In most of the business districts the zoning requirement allowed greater heights and the ratio was 1:2½.

12. Hugh Ferriss, "The New Architecture" in *The New York Times*, 19 March 1922: 54.

13. This article begins with a general criticism of the "frontal" character of New York architecture prior to the zoning amendments. It then discusses a few early explorations of the aesthetic potentials of zoning by contemporary architects. Corbett's main examples for illustrating the favorable effects of zoning are the Fisk and Liggett buildings. Harvey W. Corbett, "High Buildings on Narrow Streets," in *The American Architect*, 8 June 1921, pp. 603–619.

14. Another early source of inspiration for the architects of the 1920s can be identified as the drawings of Eliel Saarinen for the Tribune competition in Chicago. As this project was not

connected in any way to the zoning laws of New York, it could not have been seen as a direct response and solution to the aesthetic problems which had arisen because of the new laws. Nevertheless this project is cited since its resolution of building elements marks an early definition of the architecture of the period. Another important early project is the Shelton Hotel which was constructed in 1923 and was a hallmark in defining and giving a physical reality to the effects of zoning in a manner that was praised by some architects as pointing the way to a "new" and "modern" architecture. An illustration of this building appears in Ferriss's 1929 publication, *The Metropolis of Tomorrow*, p. 31. Ferriss praises this structure for its bold and successful arrangement of masses with a lack of regard for historicist styles which according to him plagued the architectural thinking of the time.

15 Robert Stern *et al.*, *New York 1930: Architecture and Urbanism between Two World Wars*, p. 508. One of the main aesthetic challenges facing the architects of the 1920s was the treatment of the stepped parts of the buildings. Prior to the legislation of the Ordinance when buildings were not required to be stepped, the differentiation of a façade into distinct sections might not have seemed an immediate problem. The zoning requirements introduced elevational distinctions which had to be reconciled. Reflecting on the current state of architectural design, and the new aesthetics of zoning, the architect and critic Aymar Embury II states: "we are having a series of buildings set one upon another, rather than single buildings decreasing in size as they mount . . . so far none of our tall buildings exhibits quite the continuous growth from base to summit that an architect would like to see." Aymar Embury II, "New York's New Architecture: The Effect of the Zoning Law on High Buildings" in *Architectural Forum*, October 1921, p. 124. Irving Pond also speaks to the same issue: "Tower and building are one, and the one factor is not to be distinguished from the other." Irving Pond, "Zoning and the Architecture of High Buildings" in *Architectural Forum*, October 1921, p. 134.

16 These drawings did not address the problem of the tower which, had it originated from the 25 percent lot size would have proved uneconomical.

17 As Carol Willis notes, "the striking contrast between these purely analytical descriptions of the zoning envelope and Ferriss's dramatic chiaroscuro renderings of the setback masses strongly suggests that Corbett's perception of the formal ramifications of the zoning law matured considerably after – and through – his collaboration with Ferriss." Carol Willis, "Zoning and *Zeitgeist*: The Skyscraper City in the 1920s" in *JSAH*, March 1986, p. 53.

18 Carol Willis, "Drawing Towards Metropolis" in *Metropolis of Tomorrow*, p. 157.

19 Ibid.

20 Ferriss, "The New Architecture" in *The New York Times*, 19 March 1922: 53.

21 Ferriss, *The Metropolis of Tomorrow*, p. 72.

22 It is noteworthy that Ferriss only recognizes the contribution of the architect at the second stage in relation to his role of providing light into the building. According to Ferriss, "the first step which is taken by the architect is to cut into the mass to admit light into the interiors." Hugh Ferriss, *The Metropolis of Tomorrow*, p. 74.

23 Ferriss, *The Metropolis of Tomorrow*, p. 78.

24 Evidence of the struggles of architects against the demands of clients and restrictions of the laws came about in their writings. Carol Willis identifies the early 1920s as a period that is particularly rich in the quantity of articles written and published by practicing architects on these issues. *Form Follows Finance: Skyscrapers and Skylines in New York and Chicago*, p. 79.

25 Ferriss, *The Metropolis of Tomorrow*, p. 76.

26 Ibid., p. 76.

27 This study represents a large group effort whose members included architects, engineers, contractors and building managers. Substantiating its evidence through a rigorous analysis of statistical data, many factors including the lot size, value of land, legal restrictions, and efficiency of architectural design influence the ratio of cost to net return on investments.

According to the text: "The true economic height of a structure is that height which will secure the maximum ultimate return on total investment (including land) within the reasonable useful life of the structure under appropriate conditions of architectural design, efficiency of layout, light and air, 'neighborly conduct', street approaches and utility services." W. C. Clark and J. L. Kingston, *The Skyscraper: A Study in the Economic Height of Modern Office Buildings*, p. 9.

28 W. C. Clark and J. L. Kingston, *The Skyscraper: A Study in the Economic Height of Modern Office Buildings*, p. 26.

29 A seventy-five story building defines the maximum height, while eight stories set the minimum limit. This study reveals that the income becomes zero at the height of 131 stories.

30 The expenses involved for erecting a large building as well as the prospect of buying several lots certainly necessitated a much greater investment than was necessary for a small site. Just as a large development could increase an investor's profit, the risk factor involved in case of a large vacancy rate also threatened a serious economic disaster.

31 Ferriss, *The Metropolis of Tomorrow*, p. 88.

32 Ferriss, "The New Architecture" in *The New York Times*, 19 March 1922: 53.

33 R. L. Duffus, *"The Metropolis of Tomorrow: Mr. Ferriss considers the Problem Created by the Skyscraper"* in *The New York Times,* 8 December 1929: BR6.

34 Harvey W. Corbett, "New York in 1999 – Five Predictions: Architects and City Planners Look into the Crystal Ball and Tell What They See" in *The New York Times*, 6 February 1949: SM18.

35 Ibid.

36 Hugh Ferriss's drawing for the advertisement "Does It Pay to Look into the Future? What Will Happen to New York Real Estate in the Next Hundred Years?" in *The New York Times,* 8 March 1923: 10.

37 Carol Willis, "Drawing Towards Metropolis" in *Metropolis of Tomorrow*, p. 158.

38 Ibid.

39 Carol Willis, "Zoning and *Zeitgeist*: The Skyscraper City in the 1920s" in *JSAH*, March 1986, p. 58.

40 "Hugh Ferriss, the most influential architectural illustrator of the period" quoted by David Dunlap, "Commercial Real Estate; Turning Radiator Building into a Boutique Hotel" in *The New York Times*, 11 August 1999: 6. "[The] rise of influential figures such as Ludwig Hilberseimer, Le Corbusier and Hugh Ferriss whose ideas changed forever the ways we look at cities" quoted by Ann Marie Adams and Pieter Siljkes, "Architects are Heroes at Metropolis; Exhibit Features All-star Guest List of 20th-century Designers" in *The Gazette* (Montreal, Quebec), 13 July 1991: J2.

41 *The Nonist*, "Hugh Ferris: Delineator of Gotham," 2003 in *The Nonist*, http://thenonist.com/index.php/thenonist/permalink/hugh_ ferriss_delineator_of_gotham/ (accessed 22 May 2008).

42 Steven Litt, "Glorious Renewal Looms for Shabby Police Building" in *Plain Dealer* (Cleveland, Ohio), 7 March 1993: 3H.

43 "The New York Chapter of the American Institute of Architects yesterday elected as its new president Hugh Ferriss, internationally known consultant, to succeed Francis Keally. Mr. Ferriss served as consultant to the United Nations Headquarters Planning Staff and at present is consultant for planning the Inter-American Center in Miami, Fla." Quoted in "Consultant is Elected Head of Architects" in *The New York Times*, 5 June 1952: 54.

44 "Ferriss Heads Architects; Portrayer of City of the Future Installed by League Here" in *The New York Times*, 7 May 1943: 16.

45 Complaints about the lack of clarity in the Ordinance were commonplace in its early stages of development. Even Herbert S. Swan, the executive secretary of New York's zoning committee believed that: "viewed simply as a piece of bill drafting, completely ignoring the

wisdom or unwisdom of its substance, the ordinance is so involved and complicated in its form and method of statement as to its render its meaning obscure and sometime unintelligible. This comment may probably not be altogether unwarranted for it is not an infrequent occurrence to find people who differ fundamentally in their interpretation of the ordinance, and upon occasions to find some persons interpreting the ordinance in radically different ways at different times." Thomas Hastings, "The Zoning Regulations in New York" in *Architectural Forum*, October 1921, p. 125.

46 Carol Willis, "Zoning and *Zeitgeist*: The Skyscraper City in the 1920s" in *JSAH*, March 1986, p. 49.

47 Establishing the unanimity with which the architects and planners of the 1920s celebrated the introduction of the zoning laws, Carol Willis says, "although *zeitgeist* characterizations of history are suspect in current scholarship, the pronouncements of epochal significance with which so many of the writers of the 1920s celebrated the principle of zoning are difficult to ignore or discount." Carol Willis, "Zoning and *Zeitgeist*: The Skyscraper City in the 1920s" in *JSAH*, March 1986, p. 49.

48 For example, the familiarity of the architects with the works of Le Corbusier.

49 The positive character of the early articles can be seen to decline towards the middle of the 1920s as more buildings are constructed, and as the critics feel that many of the old problems are still persisting.

50 George MacAdam. "Vision of New York That May Be: A Forecast of Manhattan Transfigured by the Zoning Law" in *The New York Times*, 25 May 1924: SM2.

51 For a further discussion of the aesthetic qualities of the zoning and its effects on architecture, see "Zoning and Aesthetics" and "Scaffolding the Sky" in Robert Stern *et al.*, *New York 1930: Architecture and Urbanism between Two World Wars*, pp. 34–35 and pp. 507–537.

52 Muriel Ciolkowska, "Hugh Ferris and the Zoning Laws of New York" in *Architectural Review*, November 1925, p.173.

53 Ibid.

54 That Ford's drawings were inadequate sources for explaining their subject is also proven by considering that shortly after Corbett used them, he was impelled to hire Ferriss to make improvements. See discussion on Ford's drawings earlier in this chapter.

55 It is interesting to note that the meanings of these four categories pose them as diametric opposites to one another. However, it is the tension between these opposing qualifications which infuses the drawings with their great visual and communicative effects.

56 Ferriss confronts the renderer with two questions: "The first is to grasp the *what* of the nature of the architectural subject to be rendered, to so ponder it as to exclude non-essentials. The second is *how* to employ the various devices of draughtsmanship so as to communicate this realization to others." Hugh Ferriss, "Rendering, Architectural" in *The Encyclopedia Britannica,* 14th Edition, 1929, p. 147.

57 See discussions earlier in this chapter.

58 Hugh Ferriss. *Encyclopedia Britannica*, 1929, p. 147.

59 Ibid.

60 Ibid., p. 149.

61 Ibid., Plate III, p. 148.

62 Ibid, p. 148.

63 Ibid, p. 147.

64 Ibid., p. 148.

65 Ibid.

66 Hugh Ferriss, "Truth in Architectural Rendering" in *Journal of the American Institute of Architects*, 13 March 1925, p. 101.

67 Ferriss, *Encyclopedia Britannica*, 1929, p. 147.

68 Ibid.

69 Ibid. The next three objectives remained for him for future development and included: "to serve as criterion and guide in city planning, to assist in evolving new types of architecture and to strengthen the psychological influence of architecture on human values."
70 Ibid, p. 146.
71 James Sanders, "Art/Architecture: Taking the Memorial Designs for a Test Drive" in *The New York Times*, 30 November 2003: 40.
72 These exhibitions include: Ferriss and Corbett's *1925 Titan City Exhibition* held at the New York John Wanamaker department store in 1925 and The Whitney Museum of American Art, May 1942. (Edward A. Jewell, "America's Power Portrayed in Art; Drawings of Great Buildings by Hugh Ferriss Shown at the Whitney Museum" in *The New York Times*, 5 May 1942: 16.) *The Metropolitan Museum of Art: An Architectural History*. Metropolitan Museum of Art, New York City, 1996.
73 "Hugh Ferriss, 72, Architect Here: Farseeing Designer Is Dead – Foe of Skyscrapers" in *The New York Times*, 30 January 1962: 29.
74 Willis, "*Zoning and Zeitgeist*: The Skyscraper City in the 1920s" in *Journal of the Society of Architectural Historians*, March 1986, pp. 54–55.
75 Ibid., p. 57.
76 Claude Bragdon, "Skyscrapers" in *The American Mercury*, 22, 27 March 1931, p. 293.
77 Ibid.
78 Ibid., pp. 294–295.
79 Ferriss, *Metropolis of Tomorrow*, p. 78.
80 Ibid, p. 60.
81 Ibid, p. 140.
82 Paul Goldberger, "Architecture: Renderings of Skyscrapers by Ferriss" in *The New York Times*, 24 June 1986: 13.

2 Graphic Integrity and Mapping Complexity

The Works of Lynch, Wurman and Tufte

The quest for "good" urban form is an ongoing concern for urban designers. Through these efforts, they have searched for concepts of "ideal space," which can be regarded as ideal in terms of form as well as in terms of satisfying human desires to understand the various characteristics of the city. This chapter focuses on the role of visual representation by graphical means for the purpose of investigating peoples' perception of the urban realm, as well as its methods and roles in explaining urban information. Furthermore, the chapter investigates the use of diagramming and mapping as a means of simplifying the complexity of urban flux (changes in urban form, i.e. the development of parks, streetscape, new buildings, etc.), in essence to reveal the complexity of the city. The role of information graphics is vital in somehow staging the complexities of the city in a visually simulating manner. Key theorists of the mid- to late twentieth century who have dealt with these issues include Kevin Lynch, Richard Saul Wurman and Edward Tufte. They were key players in such professions relating to urban design and urban mapping. The first part of the chapter focuses on Lynch's theory of urban form as well as the methods he used, such as cognitive mapping, for proving the theory that he first presented in the 1960s. In particular, Lynch used a specific communicative style to "expose" the "physical" elements that affect our city – "the path," "the edges," " the districts," "the nodes" and "the landmarks."[1] Lynch was a key individual in explaining through mapping and diagramming what was really going on experientially in the city, not just what a designer intellectually supposed. Lynch's book, *The Image of the City* (1960), inaugurated a new science of human perception and behavior in relation to the city, and is still used in architecture and planning courses as a key resource. The second part of the chapter examines the importance of visually communicating graphics in a truthful and telling manner, with references to the well-known information architect

Richard Saul Wurman and to Edward Tufte, Yale University professor emeritus of information graphics and statics. Wurman offers a cognitive organization of geographic information, using mapping strategies as a means to understand urban information. In 1984, he established the famous TED conferences that brought together creative thinkers in the fields of technology information and design. Experts in the field of mapping visualization, including Hans Rosling (who became a regular at TED after 2003) and Stephan Van Dam, presented their research work at these conferences. Finally, the chapter considers Tufte's work and his lifetime goal of visualizing information. For Tufte, the act of arranging information ultimately becomes an act of insight.[2] His books on information graphics capture a wide range of audiences from the architect to individuals in data and information management. We will look at the recent and current work involving methods of drawing and mapping in order to solve or understand urban information.

Kevin Lynch

In 1960, Kevin John Lynch (1918–1984) profiled three American cities, Los Angeles, Boston and Jersey City, using surveys and diagrams as part of his method.[3] He recruited subjects who were mainly middle-class professionals (planners, engineers, architects) and the general public who were familiar with their environment. He asked the subjects to draw sketches of their city based on memory. He identified common elements in these maps. He later defined appropriate symbols to represent these elements, and then synthesized the many subjects' maps into a "main map" for each city. In an attempt to understand the image of a city as it emerged in the perception of its users, Lynch analyzed each urban dweller's diagram of the city he or she inhabited. The notion of *imageability* of the city was very important to Lynch's theory. *Imageability* is the quality embodied in a physical object that gives it a high likelihood of generating a strong image within a given observer.[4] How easy is it for the user to diagram his or her city? Is a city that is easily mapped a better city?

Lynch's office interview covered a series of questions.[5] One main task was to draw a quick map of the city – in the case of Boston, the area inward or downtown from Massachusetts Avenue. The subject had to draw this map as if he or she were making a quick visual description (a rough sketch) of the city to a stranger, covering all the main features. The interview lasted approximately one and a half hours. As part of his method, the subjects "were taken out in the field to go through one of the earlier imaginary trips: that from Massachusetts General Hospital to South Station."[6] Six standard trips were selected – to Commonwealth Avenue, the corner of Summer and Washington Streets, Scollay Square, the John Hancock Building, Louisburg Square, and the Public Garden.[7] Also, Lynch selected five standard points of origin including the main entrance of the Massachusetts General Hospital, the Old North Church in the North End, the corner of Columbus Ave and Warren Street, the South Station, and Arlington Square.[8] During the subjects' walking trips, an accompanied interviewer (from Lynch's research team) asked the subject a series of questions which were recorded on a portable tape recorder. Figure 2.1 is Lynch's synthesized map of Boston as derived from the oral interviews. Figure 2.2 depicts Lynch's synthesized map of Boston derived from the sketch maps. Figure 2.1 (the oral depiction) includes more detail and information compared with that of Figure 2.2 (the subject sketch map synthesis). Major elements rarely appear in only one source. Generally, what is not given a visual representation was often referred to in oral terms (or textual notes from the subjects). This observation was most apparent in the case of the Jersey City drawings. The listing of distinctive features proved to have the highest flexibility of all the measures, excluding many elements that appeared on the sketches, which highlighted visual predominance. According to Lynch, well-defined streets are recognized by over 90 percent of the interviewed subjects.[9]

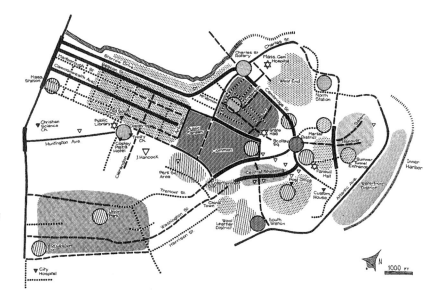

2.1
Lynch's mapping synthesis of the Boston image as derived from the subjects' oral interviews.

Source: Kevin Lynch, *The Image of the City*, p. 146, © 1960 Massachusetts Institute of Technology, by permission of the MIT Press

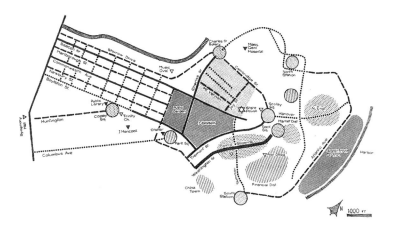

2.2
Lynch's mapping synthesis of the Boston image as derived from the subjects' sketch maps.

Source: Kevin Lynch, *The Image of the City*, p. 146, © 1960 Massachusetts Institute of Technology, by permission of the MIT Press

Ultimately, the subjects understood their surroundings in ways that are predictable and consistent and, in doing so, form mental maps, which contain five elements.[10] "Paths" include the streets, sidewalks or other channels that people make use of in order to travel. "Edges" are the linear elements that are not used or considered as paths by observers. These include walls, buildings and shorelines and are perceived as boundaries. "Districts" are the medium to large sections of a city that are distinguished by a particular character or identity. "Nodes" are the focal points and intersections, which include junctions, crossings or convergence of paths as well as places where there is a transition. Finally, "landmarks" are objects that are readily identifiable and that serve as reference points for observers, such as buildings, signs, stores or mountains. Lynch found that paths, which are identifiable, continuous and have directional quality, were the most dominant city elements for most of the people who were interviewed. This prominence is linked to the concentration of special uses or activities along a street, as well as the extent to which special façade characteristics define a path's identity. However, their importance varied with the individual's degree of familiarity with the city.[11] Edges were stronger in Boston and Jersey City, which included the Charles River and the Hudson River respectively, and weaker in Los Angeles.[12] Though Boston possesses a number of differentiated districts, which people felt made up for the city's rather confusing path pattern, and Jersey City has ethnic and class districts that have little physical distinction, Los Angeles, aside from the Civic Center area, was found to be lacking in strong regions.[13] Major transit stations as well as areas of thematic concentration, such as Pershing Square in Los Angeles and Louisburg Square in Boston, were identified as nodes.[14] Landmarks were established by their spatial prominence, either by making the element visible from many locations, such as Boston's John Hancock Building or the Richfield Oil Building in Los Angeles, or by forming a contrast with elements located nearby.[15]

Spatial pattern

Lynch proposed that three components can be used in order to analyze environmental images. These were "identity," "structure" and "meaning." Identity involves the extent to which an object is recognized as a separate entity making it distinct from others. Structure is the spatial or pattern relation that the object has to other objects as well as the observer. Meaning is a relation, though different from spatial or pattern relation, that reflects the practical or emotional sentiment that the observer has for the object.[16]

Based on the results of his interviews and analyses of the sketches made by the subjects, Lynch redrew the maps based on the subjects' testimony. Figures 2.3, 2.4 and 2.5 are images that were derived in this way for Boston, Jersey City and Los Angeles respectively. The symbols created by Lynch were used to represent the same elements for each map. Lynch drew two sets of symbols for each element; one set depicted major elements and the other minor. A major path was represented

2.3

Lynch's mapping synthesis of Boston, depicting Boston's visual form as seen in the field. The symbols, devised by Lynch, represent the five elements: path, edge, node, district and landmark (as major or minor).

Source: Kevin Lynch, *The Image of the City*, p. 19, © 1960 Massachusetts Institute of Technology, by permission of the MIT Press

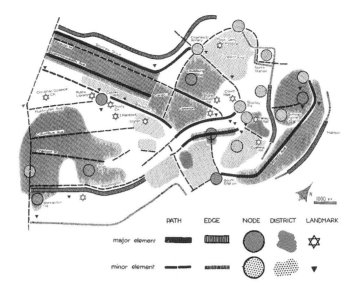

2.4

Lynch's mapping synthesis of Jersey City, depicting its visual form as seen in the field.

Source: Kevin Lynch, *The Image of the City*, p. 27, © 1960 Massachusetts Institute of Technology, by permission of the MIT Press

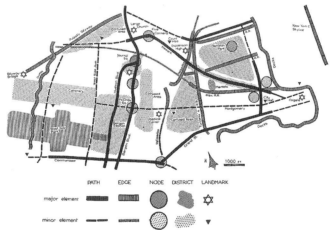

2.5

Lynch's mapping synthesis of Los Angeles, depicting its visual form as seen in the field.

Source: Kevin Lynch, *The Image of the City*, p. 33, © 1960 Massachusetts Institute of Technology, by permission of the MIT Press

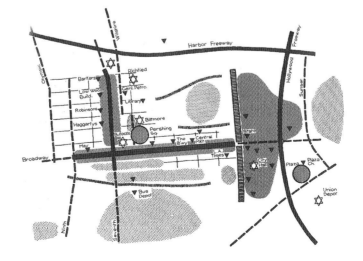

by a thick black line, whereas thinner dashed black lines depicted minor paths. A thick band (framed by parallel thin lines and filled with a perpendicular line hatch) represented a major edge, whereas a similar thinner band (unframed) represented a minor edge. Nodes are usually represented by circles or asterisks. In this case, a major node is depicted by a hatched filled circle, while a minor node is depicted by a lighter dot-filled circle. An unframed hatch pattern (with the same intensity as the hatch pattern within the circle) symbolizes a major district. An unframed dotted pattern (with the same intensity as in the minor node circle) symbolizes a minor district. Six-pointed stars represent major landmarks, whereas inverted triangles (black filled) represent minor landmarks. For comparative purposes, the symbols used to demark paths, edges, nodes, districts, and landmarks are standardized in the images for all three cities of Boston, Jersey City and Los Angeles in Figures 2.3, 2.4 and 2.5. These images are Lynch's composites of the responses from many individuals. Lynch developed the symbols to represent each element. For example, the subjects drew lines to represent roads and other paths; quickly scribbled circles would represent points of interest or intersection; and an "X" or "∗" would depict a landmark. Lynch stylized these symbols to represent the five elements he developed as criteria for the mental map composites. The same scale is also used for all three maps. The results of the field analysis for the Los Angeles and Boston areas proved to give precise predictions of the images derived from the oral interview material. Lynch's interviews were conducted to test his theory of *imageability* to gain an approximation to the public images of Boston, Jersey City and Los Angeles, and to develop a quick method for extracting or drawing out the public image of any given city.[17]

In total, paths, edges, districts, nodes and landmarks are the building blocks used to make firm and differentiated visual structures within the urban environment.[18] Lynch believed they are elements by which the city is conceived. A gathering of people in a plaza or square forms a node. Districts are created by neighborhoods, parks or areas. "Every city has its intimate inner pattern: the streets, squares, and other openings that make buildings accessible and livable."[19] This, of course, suggests a number of things about how each of these elements can be utilized, and the relations between elements, as well as how people perceive each of the three subject cities. For example, significant aspects of paths include the visual hierarchy of streets and routes as well as clarity of direction in the line of motion.[20] Also, landmarks do not necessarily have to be large objects, but their locations are crucial. More specifically, if a landmark is large or tall, then the spatial setting must allow people to see it, whereas if it is small, then it is important that certain zones, such as floor surfaces or nearby façades, at or slightly below eye-level, allow it to receive more perceptual attention than other objects.[21]

In general, these elements act both independently and in combination. Paths consist of any horizontal physical line of movement. In diagrams, roads are often expressed as simple lines or curves when describing a city. A route is one of the easiest entities to reference visually. Edges can be further defined as barriers between one section of the city and another, such as shorelines, walls, or edges of different developments. Usually the urban dweller does not represent this type of element as often as roads when describing the city. A road is one of the simplest and most easily recognizable elements to express visual character in the form of diagrams. Districts are identified as large areas with common characters or features, such as Central Park in New York or the Yorkville area in Toronto. Districts are often mapped as blobs or shaded rectangular spaces, and are usually framed or defined by edges. Nodes are often visually expressed as circles, dots or asterisks. The common city dweller does not usually express this type of information as readily as an urban designer. Architects or urban designers used the node as a means to represent a point of change, interest, intensity or junction. Furthermore, landscape features of the cities, such as water or vegetation, were often noted with pleasure and care.[22] This is how Lynch's subjects chose to represent such features in their own maps, as opposed to the standard symbols that Lynch himself adopted, as in Figure 2.3. Overall, this suggests something important about cities, and how people interact with these elements as well as with their urban form and planning:

> We are continuously engaged in the attempt to organize our surroundings, to structure and identify them. Various environments are more or less amenable to such treatment. When reshaping cities it should be possible to give them a form which facilitates these organizing efforts rather than frustrates them.[23]

A large part of Lynch's theory is rooted in the apparent visual quality or "legibility" of the cityscape.[24] What is the overall pattern of the city? How easily can this pattern be mapped? These are some of the concerns that shaped Lynch's theory of urban form. Visual sensations such as color, shapes, lights, size and scale are all components for orientation. As parts of a wayfinding system, these visual cues are critical components in the *imageability* of the city. This almost becomes a way for understanding the structure of the city and the development of an environmental image (a person's own image of the city). What are a person's immediate responses to, or thoughts of past experiences of this place? The need to familiarize and map our surroundings is crucial and has such long roots in the past that this environmental image has "practical and emotional" importance to the individual.[25] This notion is a strong part of Lynch's theory and research. Lynch was interested in the functional and practical aspects of the mental representation.[26] This concept of "imageability" also referred to the capacity of urban elements to imprint the observers' mental map with a vivid image.[27] The identity or cultural significance of the image was not the main focus of Lynch's investigation.[28]

How an individual perceives his/her city can be mapped as a mental image. Cities that can generate clear notions of landmarks, edges, districts, paths and nodes can be diagrammed with appropriate symbols. In Lynch's research, the idea of mental mapping derives from how his subject perceives the cities using mapping as a technique to describe their perceptions. One of Lynch's primary concerns is with peoples' perception and understanding of their city's form, as well as what this form actually means to them at a practical level. In the creation of the simple plan view maps, Lynch clearly defined these key five elements that "made-up" the city. The circles, the asterisks, the stars, thick line versus thin line to demarcate primary and secondary elements – these markings become part of Lynch's stylistic and communicative measures in depicting these elements.

However, Lynch's "static" diagrams/maps fail to address issues of urban flux, i.e. circulation patterns (vehicular and pedestrian), circulation volume, crime indices and other changing social and demographic parameters. These are other influences that affect *imageability*. Though Lynch touches upon these issues, they are not a prime concern in his research. As part of the sketch surveys of the cities, Lynch found that many of his subjects mapped Los Angeles easily due to its grid system. The Los Angeles grid provided an automatic alignment between downtown streets. It was easy to draw such a basic pattern in a sketch map. Two-thirds of his subjects drew this feature first before adding any other elements.[29] In a later survey taken in New York, respondents would draw a grid system of the road network first before adding in any other elements or buildings. New York City's grid pattern is a powerful image that a person recognizes when he or she thinks of this city. People's perception or constructed image of the city, according to Lynch, is based on the Gestalt principles of cognitive organization, which takes fragments or parts of the city and organizes them in a hierarchy of configurations and remembered experiences.[30] Overall, Lynch found that the way that an individual perceives the city results from an interactive process.

Cognitive mapping

More specifically, perception is based upon what the observer sees and how he or she interprets and analyzes their city. What captures their attention? Do any of the elements within the city evoke an emotional response? Forming this mental image is the process of mental mapping or cognitive mapping. A person's perception of their city or a specific place can be described as a mental map. A mental map is an individual's own map of their known city. Mental maps of individuals can be investigated by asking specific questions based on qualitative reference. During Lynch's experiments, in which he asked people specific questions about their lived environment, he would ask individuals for directions to a landmark or some other location, and ask the person to draw a sketch map of their city or parts of their city, in addition to what they recalled. A central issue that Lynch examined was how people organize or structure their surroundings mentally as well as how they begin to identify them visually. His research examined the visual quality of the city. This reflects the ease with which its parts can be recognized and can be organized into a coherent pattern. The objective organization is the physical geometry and structural pattern of the city. These are some of the concerns that form part of his research and theory of urban form. New York City's structural pattern is based on a grid system with Central Park located in the middle of the city. This park is a landmark for the city. Other key landmarks (such as The Empire State Building), key paths or roads (such as 5th Avenue), districts (such as Upper East Side) and nodes (such as Times Square) are the physical elements or the objects of the city that collectively help define the city as a visual whole. By this I mean these objects become the urban ingredients through which one understands or begins to perceive the city as a subject matter and create a mental image of the place. How are the elements of the city positioned, and what is their spatial hierarchy? Answers to these questions play a vital role in an individual's perception of the city. These questions become the potential framework for an image of the city. Respondents begin to associate major landmarks in the city as points of reference, e.g. the Charles River in Boston. For example, many tourists visiting the city of Toronto use the CN Tower as a reference point for navigation in the downtown area. Lynch claims that his experiments, summarized in his book *The Image of the City*, have to do with

> the look of cities, and whether this look is of any importance, and whether it can be changed. The urban landscape, among its many roles, is something to be seen, to be remembered, and to delight in. . . . Looking at cities can give a special pleasure, however commonplace the sight may be. Like a piece of architecture, the city is a construction in space, but of a vast scale, . . . perceived only in the course of long spans of time. . . . Nothing is experienced by itself, but always in relation to its surroundings, the sequences of events leading up to it, the memory of past experiences. . . . Every citizen has had long associations with some part of his city, and his image is soaked in memories and meanings.[31]

Referring to Gestalt principles of perception, the viewer is able to retain fragmented images of the city. Usually these elements become predominant within the image of a city, such as notable landmarks or key nodes. In general, Lynch used two key methods to explore the basic concept of *imageability*. One of these was the interview process, which used a small group of citizens with some regard to their view of their environment. The second looked at a systematic "examination of the environmental image evoked in trained observers in the field."[32] According to Lynch, a city that can be easily mapped is a good city. However, as within any type of mapping system, there is some level of abstraction.

Lynch's own maps synthesized from the drawings of many respondents were not precise nor were they a true representation of reality. They were created by eliminating, reducing, arranging, rearranging, adding and subtracting parts. However distorted the maps, the images remained broadly true to the topology of the city.[33] There is a complex set of relations between three entities: the city itself; a Lynch-type map of the city made by an individual; and that individual's mental image of the city. The individual's mental images are of course different from the city itself. As mentioned previously, the mental image is an interpretation of the city; it will abstract, neglect, select and symbolize, in essence distorting reality. The real maps made by Lynch's respondents seemed stretched, with roads twisted or pulled, often changed in scale, but the sequence of road layouts was usually correct. It seems as though the road order was a permanent fixture in the mind of these individuals. Sometimes individuals formulated strong mental images of buildings that included color, shape, form and material, and at other times building images were abstract and noted by their relative location on a street. Some mental images could be described by their structural organization. Some elements seemed to stand alone with gaps and disconnected elements. This usually involved disjointed places, places that stood alone, such as predominant landmarks. In other drawings, some elements were in juxtaposition to each other. Some images included landmarks that were placed in close proximity to each other while other drawings described landmarks on a relative path. But mostly the image of the urban structure (the physical layout of the city) was quite flexible. Elements were linked as if by rubber bands. For example, a park might be connected to the corner store, which is connected to the main street, which, in turn, is connected to the school – creating a number of focal points along the way. When the urban form becomes in reality more organized and buildings become interconnected, then the image structure becomes more firm. When streets are well connected with other paths to plazas and districts, and when a node is clearly defined (whether it be surrounded by key buildings or a tree-lined space or plaza), then the mental image is easier to develop.

The Lynch map is an attempt to capture the mental image. He basically polished the results from his subjects and redrew the maps with clear symbols to represent major and minor elements. These drawings, which were rendered more neatly and clearly than the ones composed by the subjects directly, would then be used for his book. As such, even the Lynch-type map (the redrawn map) differs from the reality of the mental image, which in turn, differs from the reality of the actual city itself. And of course, the Lynch-type maps differ from a conventional map of the same area. His subjects would tend to jot down what they remembered in the layout of the city, more specifically the layout's hierarchy of importance. They omitted the less important things. If, for example, there are two major paths leading to a distinctive landmark only one may be drawn. One path is clearly more valuable (to the subject/observer) than the other and the second is therefore omitted from the map. A conventional map is highly detailed and also selects and emphasizes city layouts. Obviously, given lack of time, drawing skills and detailed mental recollection, the interviewee cannot draw a fully detailed map. Priority is given to key elements that are highlighted in the subject's brain. These elements may be places that the individual sees or has experienced repeatedly, for example the subject always takes one path rather than another to get to the same destination.

The sketches provide a visual survey. They convey a subjective picture of the general character of the area but are directed by Lynch towards identifying:

- those visual qualities thought to be valuable and worth conserving;
- those qualities thought undesirable, which require changing;
- those qualities which are changing, and whether that is for the better or worse;
- those qualities which are most vulnerable to change;
- an overall exposure of distinctive features of the city.

Lynch found visual character to be linked to other sensory qualities of a place that people live in and use, and to a set of journeys by which those people move though the area, and experience it. With Lynch's method, the survey sketch maps did not use quantitative data but rather qualitative data experienced by the interviewee. Many of the sketches presented a much more intuitive level of drawing.[34] In general, the subject divided the elements into categories of major or minor significance. Major elements are those that were quite vivid, and the individual would ask himself why this element was strong or weak in identity. The observer also looked at the types of elements, how they were put together, and what gave them a strong identity. Another objective was to create a method or technique for a visual analysis of a city, which could predict the possible public image of the city as well as other characteristics or qualities. For urban designers, Lynch's innovative use of graphic notation to link abstract ideas of urban structure with human perceptual experience liberated them from the previous strictness of the physical master plan. Lynch's research helped to establish a tool for better urban design and normative design practice. By using the subjects' mental maps as a means to reveal the city's pattern or lack of it, these maps exposed the imperfection of the city's urban structure or one might say the lack of a well-defined urban design. The maps established an analytical tool to scrutinize the city's urban form. Lynch is also concerned with how we locate ourselves within the city, and how we find our way around the city. To know where we are within the city requires the individual to build up a workable image of each part.[35] Each of these images will comprise the subject's recognition of its "individuality or oneness" within the city as a whole; the subject's recognition of its spatial or pattern relationships to other parts of the city; and its meaning (both practical and emotional) for each subject/observer.[36]

Much of Lynch's analysis focuses around one fundamental question; namely "What makes a good city?" Interestingly enough, a number of years later, Lynch acknowledged why this might seem to be an unanswerable question. "Cities are too complicated, too far beyond our control, and affect too many people, who are subject to too many cultural variations, to permit any rational answer. Cities, like continents, are simply huge facts of nature, to which we must adapt."[37] But there have been many attempts in the past to document this complexity visually, whether by mapping or diagramming. What Lynch tried to do was to attempt to understand the city's form and organization through his categorized elements of paths, edges, districts, nodes, and landmarks, as well as how people understand these elements visually and mentally. These elements become determinants of the image of the city. "There seems to be a public image of any given city which is the overlap of many individual images. Or perhaps there is a series of public images, each held by some significant number of citizens."[38] These images are important for recognizing people's perceptions of the city that they inhabit. This perception is closely linked to people's understanding of their city. Lynch also helped visually show this though the simple markings and diagrammatic notations that identified these elements to the common viewer.

Lynch also believed that the idea of drawing attention to the question of what makes a good city was significant for other reasons:

> Decisions about urban policy, or the allocation of resources, or where to move, or how to build something, *must* use norms about good and bad. Short-range or long-range, broad or selfish, implicit or explicit, values are an inevitable ingredient of decision. Without some sense of the better, any action is perverse. When values lie unexamined, they are dangerous.[39]

Essentially, answering the basic question of what makes a good city not only results in an assessment of urban form and structure, but also reveals the values that underlie this particular process of assessment. Though values are an inevitable part of decision-making, there is also a need to understand and express such values. In this respect, the seemingly simple question of what makes a good city can be clearly linked to a great deal of significance as well as complexity. But Lynch also showed that this is definitely a question worth asking.

Richard Saul Wurman

Among individuals who have attempted to display urban information in a visual form, it is difficult to ignore the figure of Richard Saul Wurman (1936–) and his contributions to this area of work. As an architect, academic and graphic designer, he has spent most of his life solving urban problems through visual means and has continually displayed a passion for communicating visual information clearly.[40] Much like Lynch, Wurman finds the mapping process as a means to visually help understand elements and factors of the city. He is probably best known for his *Access* book series, such as *NYC Access* and *London Access*, where he claims to have developed easy-to-follow visual guides to those cities. In 1991, Wurman received the Kevin Lynch Award from MIT for his creation of the *Access* guides.[41] Wurman's publication *Understanding USA*, which is also accessible on the internet, was developed from his journey towards understanding urban issues or, more specifically, urban data.[42] In this work Wurman, for example, expresses population density across the USA, showing that NYC and LA are the two most populous cities, and other cities along the coast have more population than the interior states (Figure 2.6). He carefully portrays this urban data in a method that ordinary people, as well as professionals such as architects, can easily comprehend. His computer-generated and digitally altered images stand as both works of art and visual instruments intended to guide the viewer into the country's unacknowledged urban issues. The map was created using the 1990 US Census data loaded into MapInfo GIS software as a base. Wurman and his team produced a grayscale of this map. "Light tones represented high populations and dark values sparse populations."[43] The team then converted the grayscale image into a 3D model using FormZ (3D modeling software). The team applied image modification touch-ups using Adode Photoshop and drew state boundary lines and city labels using Adobe Illustrator. Wurman's team applies digital artistry to make the map not only an informative image but an artistic one. A great deal of Wurman's work has focused and built upon issues that relate to disorientation, or the unknown that he has encountered in his own life.[44] He creates maps to reveal things he otherwise did not know previously.

Understanding USA is filled with public information about the USA, from population density to health funding. Wurman has made this publication available on his website.[45] In a more general sense, his ideas stemmed from a very important force and dealt with a significant, yet fundamental, question: How could he make America more understandable to Americans? Public information deals with everything that

2.6
Richard Saul Wurman's map of the USA's population density (map drawn by Don Moyer, art direction by RSW).

Source: Richard Saul Wurman, Reed Agnew and Don Moyer, "Population and Becoming President," Richard Saul Wurman, *Understanding USA*. (2006) http://www.understandingusa.com/

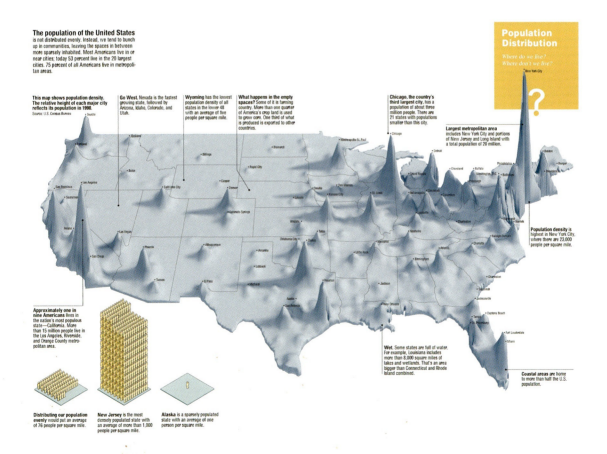

Graphic integrity and mapping complexity

we agree should be available to the political body, but making information public is somewhat less common, and difficult to achieve.[46] For information to be public it must be presented and structured in a way that is accessible and understandable to the public. Wurman takes information and attempts to make it more understandable for the public through visuals, such as diagrams or maps. His work also reflects on the concept of "information anxiety." Information anxiety is caused by the gap between what we understand and what we think we should understand. This gap is constantly widening. In short, it is the black hole that exists between data and knowledge and it occurs when information does not tell us what we want to know or what we need to know.[47] However, information is largely used to make decisions with the hope of succeeding.[48] Hence, overcoming information anxiety is crucial to the promotion of understanding. "To Wurman, content design is about creating understanding."[49]

Even though we are surrounded by information which impinges with various degrees of immediacy in our lives, these degrees can be roughly divided – according to Wurman – into five rings. The first of these is *internal information* (the messages that run our internal systems and enable our bodies to function). Although we are most affected by this level of information, we have the least control over it. Successive levels include *conversational information* (formal and informal exchanges and conversations that we have with people around us); *reference information* (information that runs the systems of our world, such as science and technology, and the reference materials that we turn to in our daily lives, such as textbooks or telephone books); *news information* (information that is transmitted by the media about people, places and events that might not directly affect our lives but that can influence our view of the world); and *cultural information* (information about history, philosophy and the arts, which is the least quantifiable form of information).[50]

With his creation of the TED (Technology Entertainment and Design) conferences, in 1984, Wurman in a way, has continually tried to look at communicative methods (maps) through new technologies, graphics and design to bring forth the "unknowns." These conferences have become vehicles to disseminate information through graphical means and have been described as a global community of remarkable people and remarkable ideas. The conferences gather individuals with a similar mission, namely the goal of making information available through new technologies and visual means.

Clearly, Wurman's work reveals that he, like Lynch, believes that urban information needs to be mapped in some visual form. Wurman met Lynch once and shares Lynch's premise of understanding information about urban conditions and making sense of them through visual documentation, i.e. mapping. Furthermore, Wurman feels that most urban information has not been presented in the past in the form of detailed maps that guide the individual to specific answers. In a direct response to this, a group of individuals, whom Wurman calls *information architects*, act to make life forces more understandable. Wurman coined the term in 1976 in reaction to a society that had reached the point of creating massive amounts of information but with little care for or order in dealing with such information. Wurman's *Access* travel guidebooks firmly established his role and responsibility as an information architect.[51] Wurman has also examined how to make information less threatening.

At its most fundamental level, the main issue that concerns Wurman is how can you empower your audience to use the information you provide them? How are you going to show information? When communicating information we do not always know that the picture in our head is the same as the pictures in other people's heads. When we instruct people, we hope that the image in their head is the same as ours, but this is not always the case. A good map can instruct people. Wurman lists three means of description: words or text, images and numbers. Generally, the best instructions are dependent on all three, but in some situations one should predominate, while the other two support. The key to providing good instructions lies in the ability to select the appropriate means.[52] Individuals are always seeking intelligent means to exchange information. Also, mapping information through images is a creative way of conveying information visually, especially if this information pertains to or is related to urban conditions or forces. In general, a map provides people with the means to share the perceptions of others; they are the metaphoric means and the tools by which we can understand and act upon information from outside sources.[53]

In conversations with Wurman,[54] several questions arose about his mapping viewpoints:

Nadia Amoroso (NA): How are your city guides organized and visually presented?

Richard Saul Wurman (RSW): Every city is different, of course. Therefore, every city is treated differently. I establish the city limits based on the core of interest. Where is the heart of that city? I erase the legal city limits and create my own based on my own experiences, and what I believe are the places of interest. My guides are therefore very personal, and created through an intuitive approach. It is ironic. I rarely receive public feedback, and my guides are designed in the first place for myself. They are personal guides, which in turn, I produce for the public. The popularity of these guides and the large number of people buying them and using them is good to know and a compliment.

NA: What makes your guides more attractive than others on the market?

RSW: My guides are simple to use, easy to navigate, as though "the guides talk to you." I did not like any of the guides out on the market at that point. They were difficult to follow and did not work for me personally. I use good judgment to produce good maps. A change in page is a smooth transition from one area to another. The *Access* guides of Tokyo, Paris, London, Rome and New York City are my best works. The *Tokyo Access* is definitely drawn from my personal experience. I use the Imperial Palace as my reference point for orientation. In my *New York Access* guide, I focus on Manhattan rather than the other Boroughs of the city. I also exclude Harlem, claiming that not many individuals like tourists visit that area. For *Hawaii Access*, I create a map for the five islands. In the *Rome Access* guide, I draw maps for individual regions including "along the Tiber, the Spanish Steps, Ancient Rome, the Vatican City and the Borgo, Piazza Navona, etc." My architectural background is a great asset to these *Access* guides. The buildings on the plan are drawn in 3D (extruded plan). The architectural history, design and creation of the places and the building are in great detail compared to other commercial guide books. I follow my interest, and maps help me understand things I once did not know. My *Access* guide maps were a special interest of mine. Always follow your interests; your work will be much more personal and stronger at the end.

NA: Your passion for architecture is clearly evident in your *Access* guide books. Do you follow or create a set of principles of making information – especially spatial and urban information – more accessible and comprehensible? Can you briefly comment on your mapping abstraction, about what you omit and what you include, about color, about exaggeration, and about how symbols work and don't work?

RSW: As part of my mapping principles, I use different colors to refer to different categories. For example, hotels may be marked in purple whereas open spaces and parks may be marked in green. This is a very important point. "When and why" to use different colors are critical issues in mapping information visually. Many cartographers are incorrect in their use of color. Cartographers have had it wrong. They use different colors to represent the same category such as temperature change. In a weather map, for example, high temperature readings are usually colored in red, which is fine. However, higher temperatures should still be shown in red but in richer intensities. Lower temperature readings should be rendered using the same color but with a moderate to lighter intensity such as lighter shade of red. Often a blue is used, which is wrong. Different colors should only express different categories. For example, the popular USA weather maps that are found in *USA Today* and many other American newspapers have many colors to represent temperature ranges across the states. For the summer months, extreme hot temperatures are colored in red; milder temperatures are marked in orange; and cooler temperatures are marked in yellow. The map should use just red (or some other color) in different shades. The same color will allow people to understand that the same type of information is repeated. This principle should be applied for maps representing population density and others. One color should be used throughout; various degrees of color intensity can symbolize more or less populous areas. Also another principle that contributes to making my guides successful is that not all information is crammed into one map. Putting everything in one map can make things too complex to understand. You want clarification. I use only the required information. The less information expressed, the more understandable it is for individuals. In creating a map, I use a pattern that is easy to understand and "visually pleasing" to read. My maps are personal guides, directing individuals to the points of interest of the city.

Wurman argues:

> You cannot perceive anything without a map. A map provides people with the means to share in the perceptions of others. It is a pattern made understandable; it is a rigorous, accountable form that follows implicit principles, rules and measures. Maps provide the comfort of knowing in that they orient us to the reality of place.[55]

Wurman's work also reflects a particular approach or outlook. He maintains that people hunt for information in terms of its nature or topic, not by its alphabetical listing. With this in mind, it is not surprising to find that *USAtlas*, another popular Wurman publication, follows this logic. People travel by car geographically, not alphabetically. For this reason, adjacent states are placed on adjacent pages. Furthermore, pages have driving time/distance grids.[56] This is not unusual; of course, many other map books apply this principle. Wurman also notes that the discovery of a structure makes accessibility possible.[57] But real and meaningful knowledge of almost anything is only useful when it is taken in relation to something else that we know.[58] This idea is captured in what Wurman refers to as "Wurman's First Law": "You can only understand something new relative to something you already understand."[59] At the same time, dividing a subject into understandable parts can be used to conquer its complexity, and this is particularly true if the parts are enhanced by the structure of the design.[60]

Wurman also claims that maps can be metaphors – tools that can allow the individual to perceive information in a different way. "Maps are the metaphoric means by which we can understand and act upon information from outside sources."[61] For example, air-pollution maps can be depicted using smoke or fog-like analogies. "Mapping implies systems of ordering and surveying, of creating correspondence, pattern and place and is a powerful form of symbol making. The map is also a metaphor for journey, and for locating ourselves in the interstices of time and space."[62] Maps can also take many different forms. For example, a CAT scan is a map of the human body whereas a GPS system may mark the individual's personal destination. Furthermore, ideas and concepts as well as physical places can be mapped.[63] They provide a visual means or a pattern that is made understandable. Maps provide the comfort of knowing in that they orient us to the reality of a place.[64] Maps have also been associated with representing visually the "invisible" elements of the city.[65] These include factors that are part of the city's environment, but not specially tangible elements such as a road or a building. Some examples of the "invisible" elements include population density, economic indicators of the city, and pollution readings. Throughout history, maps have been associated with power and politics, whether they depicted military sites, the location of hidden treasure, hunting grounds or trade routes; maps give the viewer a sense of perception and perspective on the information they are trying to disseminate. Comparable maps allow the viewer to draw different information from various sources, such as cities drawn at comparable scales. Figure-ground maps, first well-conceived in the 1748 Nolli map of Rome (by Giambattista Nolli), are a great means of visual comparison for cities, if drawn at the same scale. The figure-ground drawing is a simple black and white drawing, in which the footprints of buildings are blocked in black and everything else (streets and open spaces) remain white on the paper (or vice-versa).

Underlying this is an understanding of the ways that information can be organized. These ways are infinite and information can be organized by category (e.g. goods in a retail store that are organized by different types), time, location, or alphabetically.[66] Examples of time organization include organization based on events that happen over a fixed duration, such as presentations in museums. Examples of location include organization of information that comes from diverse sources or locales, such as the distribution of industries around the world or the different locations in the body in the study of medicine. When referring to the alphabet, examples of organization may be based on the twenty-six letters, which is an intelligent way for organizing large bodies of information, such as words in a dictionary. An atlas may have an overall organization by category, such as by type of map (e.g. city or regional) whereas information within these sections may be organized on the basis of location.[67] Though the traditional format for guidebooks involves a division of neat categories, such as restaurants, hotels and stores, each with their own chapter, Wurman's guides do not contain restaurants in one section and hotels in another section, but rather they are ordered by district. Consequently, these details in Wurman's *Access* guidebooks are jumbled together and divided by neighborhoods, which is the way cities are laid out and experienced. In this respect, the guidebooks are an attempt to mirror cities and to capture the fabric of urban life.[68]

Maps as the source to discovery

Wurman has maintained that his main struggle has been to discover the road or the pattern that leads to memory. A map is a visual document that "maps out" his journey to discovery:

> The junctures of road to road and path to path celebrate that connection. That connection is learning, and learning is remembering what you are interested in. The delight of trivia that touches on curiosity, which is the *Velcro* of learning, makes me smile. The quality of my work depends especially on the many wonderful individuals who make my thoughts and ideas even better.[69]

In a more general sense, Wurman's work reveals his commitment to filling the empty buckets of ignorance with things he understands.[70] Thoughtful and effective management and presentation of information, which includes an understanding of how information can be organized as well as how additional knowledge can build on what is already known, can go a long way towards dealing with the issues of ignorance as well as information anxiety.

Wurman definitely shaped the notion of mapping and the visual portrayal of information throughout the latter half of the twentieth century. His passion for making information understandable and exposing the invisibles of the city was inaugurated in his first book at the age of 26, *Cities: A Comparison of Form and Scale* (1963) and later in *Making the City Observable* (1971). He has made a significant contribution to the field of design, technology and information understandable through his TED conferences. He is interested in using maps not only to understand cities but for all kinds of information, from healthcare, to business relationships, to processes, to city typologies. His work has influenced many individuals in the field of visual representation and design as it relates to urban mapping and understanding information. This can be seen in the mapping visualization works of Joel Katz, Stephan Van Dam, Massimo Vignelli, and Don Moyer, to name a few.[71]

He still continues to push to the forefront urban issues in some kind of mapping form, with the collaboration of leading architects and cartographers in the fields. The next major project Wurman is producing involves mapping at a new level – the 19.20.21 project, which examines 19 cities in the world with populations over 20 million in the 21st century:

> While some say the world is flat, supercities are rising – vast, intensely urban hubs will radically redefine the world's future macroeconomic and cultural landscape. Most of the world's population right now lives and works in cities . . . the rise of supercities is defining megatrends of the 21st century. . . . No two cities in the world, or even cities within the same country, ask the same questions that result in the data that describes themselves. . . . In 1800, less than 3% of the world lived in cities . . . today more than half the people on earth now live in cities. . . . By 2050, it will be more the two-thirds of us.[72]

It is important to understand cities. The 19.20.21 project will focus on globalization patterns and explanations that will become key tools for mapping and understanding our future city. This global mapping project focuses on visually portraying key urban issues through web, publication and media sources. This would be one of Wurman's newest mapping endeavors. For more information on this latest work, please visit www.192021.org.

Edward Tufte

Like Wurman, Edward Tufte (1942–) is a key figure in advocating the importance of visually communicating information in a truthful and telling manner. For Tufte, the act of arranging information ultimately becomes an act of insight.[73] According to him, "what we see is always what we get," meaning that individuals comprehend images at face value. Therefore, what you see should be clear and understandable. Tufte has made the visual representation of information his life's work. Furthermore, he believes that the packaging of information is something that ultimately determines how much is accepted and used by other individuals.[74] He maintains therefore that a major goal for creators of graphical displays should be the achievement of graphical excellence and the promotion of graphical integrity. "The world is complex, dynamic, multi-dimensional; the paper is static, flat. How are we to represent the rich visual world of experience and measurement on mere flatland?"[75]

Information graphics: the principles

Tufte believes that graphical displays should show the data; encourage the viewer to think about the substance rather than methodology, graphic design or something else; avoid distorting what is being communicated by the data; make large data sets coherent; and present the data at several levels of detail, ranging from fine structure to a broad overview. Furthermore, graphical displays should serve a purpose, such as description, or exploration that is reasonably clear to the user.[76] Overall, Tufte regards graphical excellence, which is a matter of substance, statistics and design, as the well-designed presentation of interesting data. It consists of presenting complex ideas with clarity, precision and efficiency while always telling the truth about the data.[77] In addition to these principles, Tufte believes that graphical excellence should give the viewer the greatest number of ideas in the shortest time with the least ink in the smallest space.[78] It should tell the truth about the data it tries to portray visually.

Graphical integrity is maintained when the visual representation of the data is consistent with the numerical representation.[79] If not, then Tufte suggests that some form of graphic misrepresentation is likely to have occurred.[80] He claims that two main principles aim towards the individual perceiving the correct information through proper visual communication, and hence leading towards graphical integrity. The first principle is that: "The representation of numbers, as physically measured on the surface of the graphic itself, should be directly proportional to the numerical quantities represented."[81] The relative size of the symbol is critical. Size applies to all dimensions involved. For example, circles may be used to represent the populations of cities, such that the diameter of each circle is proportional to population. But when we look at circles, we tend to read their "sizes" by the area (two-dimensional) and not by the diameter which is one-dimensional. As the diameters are increased linearly, the areas of the circles increase by the squares of the radii, hence at a much faster rate. Thus the larger circles visually exaggerate the relative population sizes of the bigger cities. If circles are to be used, it would be much better to make the *areas* of the circles proportional to population. If the proportion is incorrect (larger or smaller) then the visual representation is presenting a lie, thus miscommunicating the correct data. The second principle is that: "The use of two (or three) varying dimensions to show one-dimensional data is a weak and inefficient technique, capable of handling only very small data sets, often with error in design and ambiguity in perception."[82] Using Tufte's principles, lines (one-dimensional) that are drawn in proportion to size of population (one-dimensional) may be a more appropriate way to depict this type of data. This may be true, but in exposing the hidden factors of the city, other measures should be examined. Should data be treated as space? How do you represent the spatial resultants of information (in particular the ones that affect our city structure)? How do we begin to register information as architects, landscape architects and city planners as spaces that have an influence on our design decisions? Tufte's principles in the case of exposing the forces that affect our city, should be taken at a fundamental level. We need to use modes of representation that architects and designers of the site can relate to, yet still hold true (as much as possible) the empirical value of the data. Perhaps the exaggeration of representing certain information may be the way to convince city planners of the parameters affecting our city, such as levels of crime, air quality, ozone levels, population settlements, etc.

All this sheds light on the possible sources of graphical integrity and sophistication as well as the various conditions that cause graphical mediocrity. Overall, a number of conditions may exist that can promote graphical mediocrity and result in the presentation of graphics that lie, make use of only the simplest designs, or fail to communicate what is contained in the data. This can perhaps be traced to the lack of skill of the illustrator, a dislike of quantitative evidence or contempt for the intelligence of the audience.

The "chartjunk"

We should not add decorative elements such as unnecessary hatching or borders or the incorrect use of multi-colors, as what Tufte calls "chartjunk"(Figure 2.7). There needs to be a careful balance between selecting the appropriate means in visually communicating information and the aesthetic means of portraying this information. In particular, when it comes down to exposing the hidden factors of the city, two-dimensional chart-like graphics are often not an appropriate choice to seduce your viewer. In the final chapter, we will look at the art and logic in exposing the hidden factors of the city through the use of multidimensional maps that resemble landscapes or spatial forms.

In his ideas about graphical integrity, Tufte agrees with Wurman's principles of portraying urban information correctly. Maps should not be cluttered with information. Not all data need to be included in the map. Maps should show a selective amount of information that needs to reveal particular answers. The effectiveness of

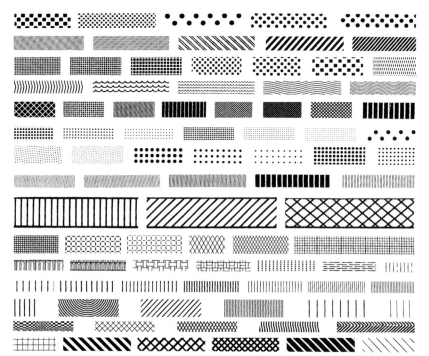

2.7
Samples of hatching patterns causing moiré effects.

Source: Reprinted by permission, Edward R. Tufte, *The Visual Display of Quantitative Information* (second edition 2001), Graphics Press LLC, p. 111

the map is a consequence of this selectivity and the careful graphical integrity of the map. Tufte concurs with Wurman's principle that different colors should represent different categories and that a change in color intensity not a change in hue should be used to represent the same category of data. Tufte not only has a great deal to say about the significance of achieving graphical excellence as well as how graphical integrity can be promoted, but also he identifies graphs and maps as powerful tools for the presentation of complex information. However, this must be done in a manner that promotes clarity, correctness and coherence. This, of course, touches on the idea that while graphical images can be an excellent means of presenting information, this is not to say that all graphical images are capable of achieving this goal. Hence, not only should an effort be made to promote graphical excellence, but professionals and users must be aware of how and why graphical integrity might be compromised. This in turn emphasizes the importance of not only creating but also examining and assessing graphical displays with a critical eye.

Clearly the works of Kevin Lynch, Richard Saul Wurman and Edward Tufte reflect important similarities and differences. While all three are concerned about the information that can be presented through graphical images such as maps, both Wurman and Tufte focus a great deal of effort on how information might be communicated more effectively to users in order for them to better understand various topics of concern, such as cities. Lynch's seminal work dealt with the impressions that cities left on their inhabitants as well as what was revealed about these cities through what people communicated about them in their interviews and sketches.

However, all three of these theorists overlap in one important way when dealing with the topic of the role of visual representation in communicating information about cities. More specifically, diagrams and maps are indeed able to simplify the complexities of urban statistics, but they must always be used with a certain amount of care and caution.

Through his work, Lynch showed that maps and sketches were able to not only communicate rich information about cities, which likely would not have been available through oral or textual expression, but they also revealed something important about the quality of urban form. In his efforts, Wurman has been able to create more effective and meaningful maps for people and, in a more general sense, show how it is possible for "information architects" to reduce the "information anxiety" that people are more likely to experience in this day and age, which, of course, is something that has become a topic of increasing importance as computerization and other digital technologies have continued to create rapid growth and change. Tufte, who has also been identified as an "information architect," is involved with work that is more closely related to what Wurman has done, and continues to do. Through his series of books including *Envisioning Information*, *Visual Explanation* and *The Visual Display of Quantitative Information* he provides detailed analyses and instructions for creating maps and other graphical images that are more meaningful to users, by avoiding problems that relate to graphical integrity, such as the presentation of chartjunk.

Maps allow people to find their way and this fact allows mapmakers to communicate information to users about a particular topic or environment, which can often be unfamiliar and/or rather complex. In this respect, maps play a useful, if not vital, function in simplifying the understanding of complex urban changes through visual means. However, it is equally important to consider the complexity that is associated with maps, which not only relates to how they can be created and presented, but also how they are often, if not invariably, embedded within a broader social, political, cultural, geographic, technological and economic environment. This, of course, urges mapmakers and users to try to achieve a deeper understanding of what maps can and do say.

Lynch took the view that paths, edges, districts, nodes and (of course) landmarks constitute the building blocks from which a differentiated visual depiction of any urban space can be conceived. Beyond that, Lynch always maintained that each city has its own intimate inner pattern and that the five components determine the rhythm, hierarchy, and foci of the urban space; in other words, the best way to really "see" the city's true nature is by looking at these items and at how they are arranged – for these things will reveal much about local power dynamics, relationships between and amongst people and between and amongst different locations, and in what ways (in what directions) the city is evolving.

In many ways, Wurman is less rewarding to discuss than Lynch because he offers information and insight that is not particularly insightful. Be that as it may, he is correct to point out that maps exist to help people order information; furthermore, they do provide three means – text, images and numbers – by which the artist can convey information to the uninformed. Needless to say, it is up to the artist or artist/cartographer to find the best means by which he or she can metaphorically communicate urban data. For his part, Tufte is adamant that maps should not be cluttered; this is wholly unacceptable in his view. Instead, he feels that data should be introduced selectively – it is not necessary to have a map that incorporates everything. The key is to devise a map that gives just enough information to answer certain pertinent questions.

While maps can deal with issues that relate to complexity they are also a source of their own complexity. Maps seemingly exist on many different levels and their visual representation can be influenced by many factors – including change over time. A mapping strategy to help simplify complexity could perhaps be that of layering and sorting intricate information through diagramming. Nonetheless, because of their multidimensional nature, maps which endeavor to illustrate longitudinal change or phenomena involving a number of different factors or variables will also be difficult things to make – and difficult things for people to understand.

The next chapter will examine the concept of "datascaping," specifically that devised by Dutch architect Winy Maas of MVRDV, and will examine mapping as it enters new modes of visual representation using digital technology. The chapter further speculates on the positive contributions that digital visual expression has made in the realm of mapping urban phenomena, and visualizing information beyond portraying the existing conditions. The chapter will conclude with an interview with Winy Maas on the topic of his datascapes and the future applications of the datascaping concepts.

Notes

1 Lynch, *The Image of the City*, p. 46.
2 Jensen, *Simplicity*, p. 140.
3 Three disparate cities of Boston, Jersey City and Los Angeles were chosen and a central area of about 2½ miles by 1½ miles was taken for the study. Thirty people were interviewed in Boston and fifteen each in Jersey City and Los Angeles for information such as descriptions, locations and sketches and also for the description of imaginary trips. Boston was selected because it was vivid in form and full of "locational" difficulties. Jersey City was selected because of its apparent formlessness and because, at first glance, it would appear to have a very low order of *imageability*. Finally, Los Angeles represents a new city that has a central area with a gridiron plan (Kevin Lynch, *The Image of the City*, pp. 14–15). The central peninsula within the line of Massachusetts Avenue, which because of its age, history and European flavor is rather unique among American cities, was the area of Boston that was selected for analysis. Though observers noted that their favorite views were often those across the Charles River, which with its bridges also made a strong edge, the city's image was characterized by confusions, floating points, breaks in continuity as well as a lack of character or differentiation (Lynch, *Image of the City*, pp. 16–25). Jersey City was defined by Journal Square, which is one of two main shopping centers, yet the maps sketched by the observers were often fragmented and contained large blank areas. Also, the city did

not have a defined center and was typically regarded as a collection of many hamlets. Furthermore, the most prominent symbol in this city was actually the sight of the New York skyline across the Hudson River (Lynch, *Image of the City*, pp. 25–29). The central business district and its fringes were examined in Los Angeles and observers regarded it as heavily charged with meaning and activity as well as having the basic pattern of a regular grid of streets (Lynch, *The Image of the City,* pp. 32–33). Descriptions of the city included references to the extent to which Los Angeles is spread out, spacious and formless (Lynch, *The Image of the City*, p. 40).

4   Lynch, *The Image of the City*, p. 9.
5   Refer to Appendix B, p. 141, Lynch, *The Image of the City* (1960).
6   Ibid., p. 142.
7   Ibid.
8   Ibid., p. 143.
9   Ibid., p. 145.
10 Lynch, *The Image of the City*, pp. 47–48.
11 Ibid., pp. 49–54.
12 Ibid., p. 62.
13 Ibid, pp. 66–67.
14 Ibid, pp. 74–75.
15 Ibid., p. 80.
16 Ibid., p. 8.
17 Ibid., p.140.
18 Ibid., p. 95.
19 Banerjee and Southworth, eds., *City Sense and City Design: Writings and Projects of Kevin Lynch*, p. 45.
20 Lynch, *The Image of the City*, p. 96.
21 Ibid., p. 101.
22 Ibid., p. 44.
23 Ibid., p. 90. Quotation by Kevin Lynch.
24 Ibid., p. 2. *Legibility* and *imageability* are sometimes used as interchangeable terms. Legibility deals with the ease with which the parts of the city or its whole can be recognized and can be organized into a coherent pattern.
25 Ibid., p. 4.
26 Willem A. Sulsters, "Mental Mapping, Viewing the Urban Landscapes of the Mind" (conference paper) in www.Tudelft.nl, p. 2.
27 Ibid.
28 Ibid.
29 Lynch, *Image of the City*, p. 57.
30 "Gestalt" is the German term for pattern, shape or configuration. The gestalt theory addresses issues of how we organize parts of our perceptual field and try to unify them as a meaningful whole. The whole vision is greater than the sum of the parts. (R. E. Smith. *Psychology*, p. 21).
31 See http://libraries.mit.edu/archives/exhibits/lynch/index.html for Lynch's thoughts.
32 Lynch, *The Image of the City*, p. 140. Some interview questions that Lynch used in the process were: What first comes to your mind? What symbolizes the word "Boston " for you? How would you broadly describe Boston in a physical sense? Draw a quick sketch map of central Boston, as though you were to describe the city to a stranger. Give complete and descriptive directions for the trip that you normally take going from home to where you work. Do you have emotional feelings about various parts of your trip? What are the most distinctive elements of central Boston to you? Other questions were asked. The interview process lasted about one and a half hours.
33 Ibid., p. 87.

34 Trained observers (architects and planners) mapped the areas' presence, visibility, and the interrelations among the landmarks, nodes, paths, edges and the districts, and marked the strength and weakness of these elements.
35 Ibid., p. 115.
36 Ibid.
37 Kevin Lynch, *A Theory of Good City Form*, p. 1.
38 Lynch, *The Image of the City*, p. 46.
39 Lynch, *A Theory of Good City Form*, p. 1.
40 Bill Jensen, *Simplicity: The New Competitive Advantage in a World of More, Better, Faster*, p. 142.
41 Richard Saul Wurman, "Introduction" in *Richard Saul Wurman's UnderstandingUSA*, http://www.understandingusa.com/wurman.html.
42 Wurman claims his books are the products of the solution of the problems he wished to resolve, not really for the people but for himself. He shares this information through his publications (personal communication, 16 April 2006).
43 Wurman, "Population & Becoming President" in *UnderstandingUSA*, http://www.understandingusa.com/chaptercc=3&cs=42.html.
44 In 1980, Wurman moved to Los Angeles and after realizing that he was disoriented and unable to find his way around he decided to produce his own guidebook, which accessed everything he wanted to know about the city. The result was *LA Access* (Richard Saul Wurman, *Information Architects*, p. 28). Beginning with *LA Access*, *Access Guides* have been published for more than thirty cities. "The book *Medical Access*, which is about diagnostic tests and surgical procedures, explains 32 common surgical procedures using a system of anatomical schematics on many layers. . . . The nature, preparation, procedure, and recovery stages of each operation are described on one tightly structured page. The book also answers common questions, explains 120 diagnostic tests, and shows an operating room." (Wurman, *Information Architects*, p. 29). The underlying reasons for doing *Medical Access* were similar for Wurman's reasons for doing *LA Access*. Wurman was due for a physical, but couldn"t find any literature that could provide him with the appropriate questions to ask his doctor and any good medical illustrations. The photographs he found on the subject matter made him ill. *Medical Access* consists of three sections, which include diagnostic tests, surgical procedures as well as questions and answers (Wurman, *Information Architects*, p. 29).
45 Wurman, http://www.understandingusa.com/.
46 Wurman, *Information Anxiety2*, p. 139.
47 Wurman, *Information Anxiety*. New York: Doubleday, 1989, p. 34.
48 Thomas H. Davenport and Laurence Prusak, *Information Ecology: Mastering the Information and Knowledge Environment*. New York: Oxford University Press, 1997, pp. 116–117.
49 Bill Jensen, *Simplicity: The New Competitive Advantage in a World of More, Better, Faster*, p. 142.
50 Wurman, *Information Anxiety2*, pp. 160–161.
51 Wikipedia, *Richard Saul Wurman*. 2006.
52 Wurman, *Information Anxiety2*, p. 109.
53 Ibid., p. 157.
54 My discussions with Wurman on the topic of mapping began in August 2006 and continue on an ongoing basis.
55 Richard Saul Wurman, *Information Anxiety: What to Do When Information Doesn't Tell You What You Need to Know*, p. 260.
56 Wurman, *Information Architects,* p. 31.
57 Thomas Davenport and Laurence Prusak, *Information Ecology: Mastering the Information and Knowledge Environment*, p. 119.
58 Richard Saul Wurman, *Making the City Observable*, p. 62.

59 Wurman, *Information Architects*, p. 23.
60 Ibid., p. 27.
61 Ibid, p.263.
62 Clark University, *On Mapping Projects*. See http://www.clarku.edu/offices/publicaffairs/news/press/2005/mapping.cfm.
63 Wurman, *Information Anxiety*, p. 263.
64 Ibid., p. 155.
65 Ibid.
66 Wurman, *Information Anxiety*, pp. 59–64.
67 Ibid., p. 62.
68 Ibid., p. 48.
69 Wurman, *Information Architects*, p. 35.
70 Ibid.
71 Some of RSW's publications are available online through his website, www.wurman.com. Joel Katz is a graphic designer and information architect specializing in way-finding, cartographic and diagrammatic explanations, and environmental graphic designs. Stephan Van Dam, of VanDam, Inc., is a well-known NYC-based map publisher and cartographic designer, who invented and patented the best-selling UNFOLDS® series of pop-up maps which are in the permanent collection of the Museum of Modern Art (MoMA) in NYC. Van Dam stated, "Richard Saul Wurman has played a significant part in my life." (August 2009). Massimo Vignelli is a well-known Italian designer based in New York City; he designed the iconic 1972 New York City subway map (of Richard Saul Wurman's NYC/ACCESS subway map, Vignelli wrote it "is the best solution by far to the complex problem of representing one of the most complicated rail networks in the world"). Don Moyer is a graphic designer and one of the founders of Thought Form Inc. Moyer stated, "I've known Richard Saul Wurman since 1967 when I was lucky enough to find myself in a class that he was teaching. He's been a significant influence on many people, including me." (August 2009).
72 In a conversation with Richard Saul Wurman (2009) and detailed information from Wurman's website, www.192021.org.
73 Jensen, *Simplicity*, p. 140.
74 Davenport and Prusak, *Information Ecology*, p. 145.
75 Edward Tufte, *Envisioning Information*, p. 9.
76 Edward Tufte, *The Visual Display of Quantitative Information*, 1997, p. 13.
77 Edward Tufte, *The Visual Display of Quantitative Information*. Second edition, 2001, p. 51.
78 Ibid.
79 Ibid., p. 55.
80 Ibid., p. 57.
81 Ibid., p. 56.
82 Ibid., p. 71.

3 The "Datascapes"
The Works of MVRDV

How have contemporary practices using three-dimensional digital modeling allowed individuals (including architects and urban dwellers) to view their city in ways that "flat," two-dimensional street maps have not? During recent years, advances in technology have allowed for greater use of digital media for developing such representations. This chapter examines the role of these techniques, including data-driven visualization as means of representing the "unacknowledged" features of the city. In particular, the chapter focuses on the "datascaping" experiments of the Dutch architecture group MVRDV (Winy Maas, Jacob van Rijs and Nathalie de Vries), especially the *Metacity/Datatown* work from the late 1990s and carried into the twenty-first century. MVRDV have sought to broaden the visual vocabulary of art and architecture as it relates to mapping urban phenomena through digital experimentation. MVRDV can be recognized as the modern originators of the concepts of the datascape. In some sense, they have brought to the forefront the idea of using data to generate alternative urban and architectural forms and as a means to help guide planners to urban design decisions. This chapter profiles the role of the datascape in relationship to merging information, artistic measures and form in order to expose the invisibles of the city. This will include, for example, how MVRDV's theoretical studies on datascapes have provided alternative ways of examining and conveying visually urban information not normally visualized in a spatial manner. This includes carbon dioxide levels emitted by automobile use, land value, waste production, water consumption, etc. During the 1990s, significant exploration of digital media in architectural representation led to the development of new techniques in mapping. The new maps became increasingly poetic in their graphic qualities. Using urban data, these architects widened the digital palette and developed them into artistic media in order to investigate urban situations, through their creativity and architectural background. As a whole, these individuals made significant contributions to the role of digital mapping in describing urban form, which is worth reviewing and investigating. Moreover, their styles and techniques have been carried forward to the twenty-first century with further development in digital communication and mapping techniques in architecture.

The datascape concept

The necessary point of departure for this analysis is to define the concept of datascape. Similar to Hugh Ferriss's visual investigations of the 1916 Ordinance, datascapes serve as an architectural language, which turns statistical description into representations of the theoretical envelopes within which built forms can be placed:

> Datascapes are visual representations of all quantifiable forces, which can have an influence on the work of the architect or are even able to determine and to steer them.[1]

Another description of datascapes is that they are "visual representations of all the visual and non-visual influences that have an impact on or regulate the work of an architect."[2]

James Corner, Chair of the Department of Landscape Architecture at the University of Pennsylvania, describes datascapes as:

> imagings constructive and suggestive of new spatial formations, . . . so "objectively" constituted (from numbers, quantities, facts, and pure data) that they have great persuasive force in the bureaucratic and management aspects of contemporary city design. They differ from the quantitative maps of conventional planning in that they image data in knowingly selective ways. They are designed not only to reveal the spatial effects of various shaping (e.g., regulatory, zoning, legal, economic, and logistical rules and conditions), but also to construct a particular eidetic argument.[3]

Furthermore, datascapes have been described as "revisions of conventional analytical and quantitative maps and charts that reveal and construct the various shape-forms of forces and processes operating across a given site."[4] Dutch architectural critic and historian Bart Lootsma argues that:

> Datascapes tend more to analysis and representation. They are visualizations of laws, rules, norms and statistical probabilities, and as such they constitute representations of what the sociologist Anthony Giddens calls "abstract systems" – that is, bureaucratic systems where the trust in the system as well as the people, institutions and machines

that represent it, lies in one's confidence in a certain specialized expertise. In reality, these Datascapes show that the space around us is virtually shot through from the outset with the dominant forces of society. In a single design, there are several abstract systems at work. These systems nevertheless indicate the maximum limits within which the architect can produce his designs.[5]

Lootsma is arguing that Maas (MVRDV) is using architectural autonomy to impose "expert" authority, that is, the architect plays a part as the director of political powers in design. Giddens's systems in the context of work of MVRDV: Datascapes are visualizations of what Giddens calls

> the "expert systems" and "abstract systems" that have replaced traditional authority in contemporary society. They are bureaucratic systems in which trust in the system is grounded in an assumed expertise in a particular field. Moreover the objectivity of the information in the various systems is debatable. Particular interest groups can hire their own experts to dispute the information used by another.[6]

The datascape is a facilitator of information. Maas's datascapes rely "upon an acceptance and assumption that the reality in which architects are attempting to operate is largely quantifiable and exceedingly complex."[7] Maas, as the architect, becomes an "expert" in portraying the "abstract systems" that surround our society through datascapes. Brett Steele, former Director of the Architectural Association in London, argues that the adaptation and modification of the architect's basic formal language (in design and in construction of living spaces) is now a reaction to current conditions in society.[8] The architect applies his or her professional, technological abilities elsewhere (besides the typical architectural design project, for example in other fields including urban planning or technology):

> A renewed attention on the part of the architect to his or her own procedures for quantifying, analyzing, and transforming the organizational and formal features of a project does more than merely reflect the growing role of data processing in all aspects of social life. Much of today's global information economy is dominated by organizations competing on the basis of their ability to acquire, manipulate, interpret and use information effectively.[9]

Revisiting *Metacity/Datatown*: the spatial map

A better understanding of the concept of datascapes is likely to be achieved by considering these descriptions or definitions in light of the work that has been done in this area by MVRDV. In 1999, MVRDV brought the concept of the datascape to the forefront with the publication *Metacity/Datatown* which is based on a theoretical premise in which MVRDV quadrupled the current population of the Netherlands and portrayed the consequences in visual statistical terms. *Metacity/Datatown* is originally an installation of video projections which was exhibited at the Stroom Center for the Visual Arts in The Hague in 1998–1999, and in other locations including Yale's School of Architecture. The first part of the book, "*Metacity,*" describes the

> endgame of the current trend toward urban expansion. A series of graphics excluding areas such as deserts, forests, and mountains defines the habitable area of the globe to about one-third of the land mass, before moving on to data profiles of Mexico City, São Paulo, and, for an interesting contrast, the Netherlands.[10]

The next part of MVRDV's publication documents visually the "*Datatown*" installation, in which Maas tests "extreme scenarios" for a high population density in the Netherlands by considering the six sectors – the living, agriculture, $(C)O_2$, energy, waste, and water consumption needs of the country, which will be examined further in this chapter. *Metacity* relates to the whole world, while *Datatown* relates to the Netherlands. *Metacity/Datatown* examines the spatial, social and economical consequences of an ever-expanding population density using statistics and shocking graphics.

Maas takes on a role that is different from the traditional architectural objective of designing buildings and sites; specifically, he wants to use his architectural talents and statistical insights to help humanity create for itself a sustainable, orderly world. If composed correctly, Maas's datascapes have the visual power to inform the reader about the global challenges facing all of us as members of one human community. Maas reveals the fundamental questions that inspire the *Metacity/Datatown* project:

> Can we understand the contemporary city at a moment when globalization has exploded its scale beyond our grasp? Have we lost control of its quantities, or can we analyze its components and manipulate them? Imagine a city that is described only by data. A city that wants to be explored only as information. A city that knows no prescribed ideology, no representation, no context. Only huge, pure data.[11]

Metacity is the whole world: Europe, Southeast Asia and the edges of North America are becoming continuous urban zones. In *Metacity*, Maas takes the total area of the earth's surface, subtracts the area of the sea, subtracts the area of the mountains, subtracts other impassable or uninhabitable regions, and arrives at the net area available for urban development and agricultural use. As their populations and mass commuting continue to increase, these regions' urban zones continue to expand. Agricultural and rural areas begin to intensify due to the lack of space and the demand for it. Maas continues this mapping exercise by subtracting the active seismic areas (earthquake zones) and deserts. He examines the world's available territories, showing that only a small percentage of the earth's total surface can currently be occupied as useable urban space for living, industry and production.[12] He says that areas such as "bodies of water, mountains, deserts, jungles, polar zones, all pose reductions of the 'usable' territory for Metacity."[13] As a consequence, do the oceans, deserts, jungles need to be colonized due to the expanding regional growth, or can there exist an intelligent way of expanding the capacity of our existing boundary?[14] The final "world settlement envelope" occupies a small area of the globe – a daunting realization of the future lack of development area. Through this datascape process, Maas presents a visual critique of a world "unable to grasp the dimensions and consequences of its own data."[15] Mostly presented in two-dimensional mapping conventions, the next section, on *Datatown*, takes a leap in regards to mapping conventions by visually documenting the urban consequences spatially and by providing a potential new urban form. MVRDV is one of the first groups to bring the concept of datascape to a new level – as maps and as potential design scenarios.

The city of designing with information

Datatown further explores the concepts of the future of the *Metacity*. *Datatown* is an attempt to draw relationships out of the Netherland's national data and portray the country as one megalopolis. *Datatown* was hypothetically defined as a city of 400 km by 400 km, or 160,000,000,000 m^2 with a scenario that increased the Netherlands' population density (people/km^2) by four times.[16] It is self-sustaining within its boundaries, by means of an internal ecosystem. *Datatown* is also based on a series of hypothetical assumptions or "what-if situations" in order to test the boundaries of what is real or far-fetched, often turning these assumptions into surrealistic depictions. Using computer visualization, Maas produces strange and dystopian images of the hypothetical city, and illustrates them as understandable visual representations – fantasy-like visuals that dramatize data which nevertheless are true (although not all the time) to the numerical facts, particularly as seen in the images derived from *Datatown*. Dramatizing the data is a way to capture the audience's attention about real urban issues. We cannot see levels of energy use, or water consumption or the consequences of the overproduction of waste, but Maas clearly brought these problems to the visual forefront with his introduction of the datascape about a decade ago.

Maas investigated: What would this *Datatown* look like? Would the availability of this statistical information make it a "useful" instrument of exploration? What images would be created from this numerical approach?

One method through which urban information (in numerical form) attests its value is by the use of "extremizing scenarios."[17] These situations include for example extremely high production or consumption in the previously mentioned six data sectors (agriculture, energy, living, water, (C)O$_2$, waste). As the population of the Netherlands increases, these data sectors of *Datatown* depict the "extreme scenario" outcomes. Thus, *Datatown* is constantly in flux and can be described as an ever-changing city based on the changing information. This information was collected, sorted and synthesized into the six sectors. The data were sorted alphabetically into a field of "barcodes" (lines of information), and later resorted to fit practical needs. These barcodes can be described as "data pods" or "blocks" that collect and store urban information in specific categories such as air quality, carbon monoxide levels, ozone levels, population density, land area, crime rate indices, etc. In *Datatown*, every datascape (image) deals with one or two of these influences, and reveals their impact on the design process by illustrating their extreme scenarios.[18] Each site, therefore, has more than one datascape.

In his first sector, the "Living Sector," Maas examined Dutch standards for residential density. Close to 242 million inhabitants of *Datatown* were assumed to live at an average of 2.43 persons per residential unit.[19] What is interesting in the "Living Sector" image, is Maas's attempt to visually capture the alarming "ever-expanding" residential density into one huge red cube that would contain all the inhabitants. He then juxtaposed millions of small free-standing houses next to the large mass to compare one container for population versus individual units (Figure 3.1).

This digital image is powerful. The background is free of colour, while the foreground contains a field of green pixels dotted with thousands of little red Lego-like houses. A huge red block towers over the whole scene. This is the horrifying image if the citizens of *Datatown* decide to live in single-family housing. Maas tested a vertical garden city scenario for the "Living Sector," which consisted of a combination of stacked living units and garden units.

Like the 1916 New York Ordinance, these towers allow light and air into a dense city. These digital renditions of green and red block-like units are still a far cry from an urban design image, yet they provide the individual with a conceptual framework for understanding the urban realm in quantitative terms. In effect, these datascapes are maps that depict the magnitude that extremely high population numbers would create for living in small land areas.

The next part of *Datatown* profiled the "Agricultural Sector" (Figure 3.2). This bizarre section interpreted food consumption scenarios for the city (i.e., the number of plots required to produce X number of poultry, weight of beef or weight of wheat for consumption by the town's inhabitants). Digital images of shared plots of land invade the Dutch skies – images of pigs were collaged into the background, creating an uncanny glimpse of Maas's future project, "Pork Consumption City" or "Pig City," which were towers created to house numbers of pigs for consumption (Figure 3.3). In the "Agricultural Sector" in *Datatown*, Maas created floating rectangular pieces of farmland that seem to appear out of a tornado-like storm. Trees were replaced by large eggs and chickens in the poultry consumption area. These images promote a sense of fantasy, giving the viewer alternative insights into the city from the perspective of agriculture. This is rather strange, mysterious and surrealistic. It gives the onlooker a sense of curiosity, while tantalizing one's sense of awe. The following

3.1
Winy Maas creates one large red cube measuring over one and half kilometers on edge to represent a massive volume that would contain all the inhabitants of *Datatown*. This juxtaposition shows the shocking situation of population growth and housing needs.

Source: Maas, *Metacity/Datatown*, p. 70

3.2
Depiction of the "Agricultural Sector."

Source: Provided by MVRDV

excerpt recounts art and architectural critic, Sven Lütticken's, experience of visiting the installation:

> To view this work, one must stand in the center of four screens placed at right angles to each other, on which computer-generated images are projected. Together these form a constantly moving panorama of *Datatown*, the model city (or *Metacity*) whose design is ruled by a set of statistics on area, population density, and so on. Engulfed by *Metacity/Datatown*, the viewer finds himself flying Peter Pan-like past conical mountains, over plains, past abstract buildings, and through forests (stacked in several layers, naturally). One is also confronted with an immensely high "curtain" made of metal tubes to which windmill blades have been attached at regular intervals.[20]

He goes on to say:

> In *Datatown*, rectangular pieces of farmland come floating through the air in a kind of apocalyptic storm, only to land in a neat grid that indicates

The "datascapes"

the different uses for which the land is intended – areas for chicken farms are indicated by enormous eggs and much smaller chickens. We appear to have entered a space where the normal rules of physics have been suspended. Thus MVRDV infuses the data sublime with the unsettling suggestion that the world is animated and that technology is the contemporary form of magic.[21]

3.3
"Pig-City." Forty-story towers would include pigpens with balconies providing sunlight, ceilings rigged with self-serve straw bales, slaughterhouses and hydroponic feed farms.

Source: Provided by MVRDV

In this case, Lütticken is referring to MVRDV's datascapes (in particular those of *Datatown*'s "Agricultural Sector") as the "data sublime."[22] Kant defines the sublime as "the name given to what is absolutely great."[23] For something to be absolutely great it must have a non-comparative magnitude, which is beyond all comparison great.[24] Many of the *Datatown* images do indeed provide a contemporary version of the "data-sublime." The "Agricultural Sector" image depicts the scary reality that this could happen if we continue on the path of high consumption and extreme population growth in the city. The visual is certainly compelling, and produces a sense of mystery. The concept of having large eggs, flying plots of land, large mounds of waste, and a wall of wind turbines is so far-fetched and hard to conceive that the datascape landscapes are sublime. *Datatown* images do have an aesthetic quality that creates a sense of awe. Relationships between the faculties of the imagination (art) and of reason (data) are present in Maas's work, combining art and surrealism and data into one. What *Metacity/Datatown* is trying to do is to relate a series of numbers that indirectly suggest urban envelopes to create new spaces. It defines spatial resultants on the basis of urban data. The projects are not actual proposals for some real town, but an abstract graphic representation of visual situations. The work speaks directly to the communication of geographically based data in an aesthetically desirable medium.

In the "Oxygen/Carbon Dioxide Sector," Maas chose to represent forest scenarios in *Datatown* that were of sufficient size to absorb the carbon dioxide produced, and to provide clean air (Figure 3.4). Given that there is insufficient ground surface in *Datatown* to grow all these trees, stacked planes (i.e., growing territories) were required. In essence, the forest becomes a CO_2 absorption machine for *Datatown*. Again, it is possible to find that surreal images produced by digital means captured glimpses of stacked forests of poplar trees inserted in place of oxygen-level data (Figure 3.5). The images are quite strange, yet they give the audience a direct visual impression of the number of trees required to sustain a healthy dense city.

3.4
Depiction of "Oxygen/Carbon Dioxide Sector."

Source: Provided by MVRDV

The "datascapes"

3.5
Depiction of "Oxygen/Carbon Dioxide Sector." Stacked forests of poplar trees inserted in place of oxygen-level data.

Source: Maas, *Metacity/Datatown*, p. 129

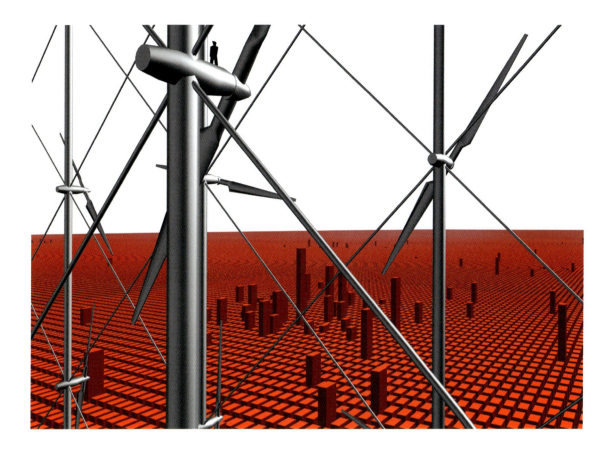

3.6
Depiction of "Energy Sector." Striking image of a windmill façade overlooking a sea of red "Living Sector" blocks.

Source: Provided by MVRDV

In the next sector, the "Energy Sector," Maas transcribed his theoretical description in a striking image of a windmill façade overlooking a sea of red "Living Sector" blocks (Figure 3.6). This image becomes the visual interpretation of the total energy demand. By carefully composing the data using digital means, Maas applied a similar approach to that used by Hugh Ferriss to reveal the unknowns and to shock the public. The sheer volume of the windmills becomes a clear indicator for the public of the enormous amount of energy needed to sustain *Datatown*.

In another sector, Maas examined the amount of waste produced daily, monthly and yearly in *Datatown*, in cubic meters. He then selected and sorted the waste into various categories, subdividing different waste products, such as household products, dredging sludge, vehicle wrecks, hazardous waste, office products and waste from construction and demolition. These amounts emerged as hills and mountainous forms, which created a new landscape (Figure 3.7). As more waste was produced, the hills increased in size and became mountains. Again, these shocking and provocative images expose the brutal truth about the future of cities if the public continues to generate vast amounts of waste – the images visualize the potential forms of the new urban conditions. Maas makes a visual argument for the need to recycle, or the mounds would eventually become mountains.

3.7
Depiction of "Waste Sector." Landscaped datascapes created by waste, such as household products, dredging sludge, vehicle wrecks, hazardous waste, office products and waste from construction and demolition. These amounts emerge as hills and mountainous forms, which create a new landscape.

Source: Provided by MVRDV

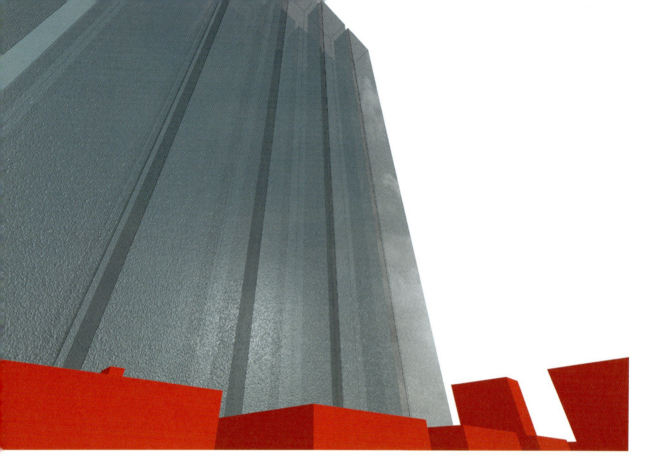

3.8
Depiction of "Water Sector," conveying large amounts of water usage. Drinking water is stored above ground in enormous rectangular towers. The sheer size of these water towers juxtaposed with the red living units gives the viewer a powerful depiction of the amount of water consumed per year.

Source: Provided by MVRDV

Maas continues to successfully conceptualize the invisible forces that shape the city in the "Water Sector." He crafts a digital rendition of large containers of water based on the ground plane of the city. Extremely large rectangular basins or towers capture the numerical amount of water the city uses per year. This visualizes the need to save our resources – otherwise these are the potential resulting urban situations (Figure 3.8).

Alternative concepts for urbanism

Maas used both artistic and statistical measures to produce the images for *Metacity/Datatown*, and he used sophisticated computer graphics to convey the information in *Datatown*. This project was conceived in 1999, a time when Maas used advanced means of digital technology to test alternative concepts of urbanism (i.e., data designing) to create overbearing and shocking images that would capture the audiences' attention. *Datatown* is "brilliantly dystopian, this is nonetheless *datascaping* in its crudest form."[25] Maas's investigations into these hypothetical urban scenarios would have been impossible to make without digital media. Using an intuitive approach, Maas renders forms three-dimensionally, giving the viewer some idea of urban data and its datascape outcome. These images take the role of maps, guiding individuals to the future urban outcomes, similar to Ferriss's Ordinance drawings. *Datatown* is a visualization of urban quantities by presenting them in volumetric forms and areas.

Maas also attempted to transform number into meaning and poetic images. The "Living Sectors" can be conceived of as communal spaces or public spaces for individuals to meet. In another image, Maas placed water from the "Water Sector" onto the "Waste Sector," and froze it to create snow for skiing and other winter activities.

The art of numbers: mapping beyond the existing

Maas tries to give artistic expression to the numerical values. Through the *Metacity/Datatown* publication, a dialectical approach is developed within the images to bring complex matters (statistical values) into a much more comprehensible form through digital art. Numbers play a critical role in analyzing these hypothetical situations, but the artistic expressions are also important in displaying these values in a way that is powerful and yet informative. Many of the images created have a fantasy-like quality to them, making them mysterious. *Metacity/Datatown* depicts dream-like fantasies or disaster situations through numbers, art and urban design. Gathering this information and transforming it into urban form is a challenge, but Maas tackles this challenge. Maas's imaginative translation of the urban data in his theoretical exploration in *Metacity/Datatown* can be summed up by the statement quoted by economist Theodore Levitt: "Data do not yield information except with the intervention of the mind. Information does not yield meaning except with the intervention of the imagination."[26]

Dutch visual artist Ronald van Tienhoven advises Maas on the power of information design:

> Information design is a new phenomenon that deals with this visual quality. An attempt is made to produce a representation of data not based on its collection or to render it as tastefully as possible while producing a matchless marriage of quantities of complex information. . . . art has a role in that process.[27]

Many of the *Datatown* images can be considered maps, since they guide the individual into future scenarios of the city based on urban phenomena that are not usually expressed in conventional maps. *Datatown* creates three-dimensional forms that provide an abstract graphic representation of urban statistics. Van Tienhoven states:

> In *Metacity/Datatown*, an attempt has been made to compile a wealth of statistical data, accompanied by a graphic or visual picture that is subsequently shown within a certain period of time in order to reflect the laws of cause and effect as clearly as possible.[28]

As part of the research to investigate the "hidden logic behind urban matter – their [MVRDV's] aim is to understand the 'gravities' that appear as 'scapes' of data behind them."[29] According to Maas:

> Datascapes are buildings that make visible aspects and opportunities of the regulatory matrix that were never intended. They emerge out of an apparently arbitrary extension of its logic, point to the arbitrariness of the rules themselves and, at the same time, produce something unexpected. The result is both rational and irrational, normal and monstrous, controlled and free, critical and affirmative.[30]

Spatial containers

Maas argues that data used to create the spatial restrictions on the area represented can be compared to abstract built forms. The datascape exposes extreme conditions and new spatial boundaries that are unexpected. Datascapes have paradoxical properties. They are rational because they derive from given data; they are irrational because the spatial restrictions produce extreme and sometimes surreal images of the area. They are normal because data is used as a method to extrapolate urban conditions; they are monstrous because the "bombastic" images produced in the process seem too extreme to imagine as real urban forms. They are controlled because the creation of the datascape uses specific criteria to abstract data sectors to certain volumes based on numbers; they are free because the datascaper uses his/her artistic hand to render the final image of the datascape. They are critical because the datascape analyzes the given site and poses a visual critique to the urban dweller, as for example if this amount of product X continues to be consumed, these are the spatial consequences; they are affirmative because data are given. The datascape process is an experimental process, using "what if" situations. What if X amount of waste is generated in the given area, what spatial consequences will occur? What extreme image is produced?

In more recent years, new modes of digital mapping have affected the ways in which space is represented visually. New techniques have allowed for more flexible means of spatial representation, as exhibited in the *Metacity/Datatown* project. Bart Lootsma believes that the datascape is "less about philosophy, theory, and aesthetics, and more about how the visionary and the pragmatic may be combined in creative and paradoxical ways."[31] He states that datascaping is concerned with "critical pragmatism."[32] Datascaping examines the practical rather than theoretical push towards suggesting and mapping the socio-ecological realities of the given site. The datascaper uses the computer as his or her tool to describe visually the changing process of the site under specific urban conditions or influences. The computer, as a tool that transforms information in a quantitative or statistical form, is a highly effective medium of visual communication. This visualization models the "complex ecological and cultural flows in relation to design interventions."[33] Within the datascape process a polarity between art and instrument merge through the hand of the datascaper and his/her computer. The tension between the two (art and data) is manifested in the "computation of datascapes as the 'cyborgian'[34] designer works within a more fluid field of data, ideas, and form."[35]

Prefiguring design

Can MVRDV's datascapes be considered maps that guide the urban designer towards better design solutions using data and the digital artistry to gain closer access to ongoing and changing urban situations? Can MVRDV abstract these concerns and highlight new conditions through the datascape map?

> A map is a construction that is simultaneously a translation of real phenomena and an abstraction of that translation. A map always exists in an uncanny relationship with the real as it is simultaneously an abstraction, a condition of opening up, of pointing towards new conditions, new spaces and new projects.[36]

If this point is true, then MVRDV's datascapes become primary maps to translate real urban phenomena or changing urban processes visually through digital means. They show extreme dystopian scenarios or extreme urban situations prescribed by urban data, and therefore, the datascape image (as a map) does present an uncanny relationship with the real urban scenario and its highly abstracted portrait. But it is these strange images that point out new conditions and new spaces.

Datascapes are more than illustrations, they reveal urban influences, which are exhibited for example when Maas takes information and transforms it into extreme images. They are multidimensional maps that guide the urban designer into potential new development and have the power to guide public policy. Within the computer, the datascape is able to adapt and take new forms based on ongoing changes in information. The datascape can also be crafted as a model using computer numerical control (CNC) outputs, then it becomes an object, or artifact, having a certain shelf life before new data replace it. As an object, that particular model holds true for the given area for a specific moment or period of time.

When assessing the development of datascapes, "The artistry lies in the use of the technique, in how things are framed and set up."[37] That is, the mapmaker takes control in showcasing the data he or she is given through creative means and artistic measures. He or she takes data and turns them into a datascape which can influence the architect or planner to think intuitively about their urban living space. The concept of the datascape promises emancipation on the basis of a comprehensive and objective collection of social requirements beyond bare assumptions or thinking habits, while opening the possibility of breaking free from existing form typologies.[38] It combines information in an interpretive manner and aesthetically powerful way. This is what makes datascapes so compelling and distances them from traditional mapping techniques used by planners. They are not tables, nor charts, nor flat color-coded diagrams, but virtual spatial forms. Datascapes serve as a fresh approach in convincing planners and urban designers to take a new approach to urban form. "Unlike the assumed and passive neutrality of traditional data maps, datascapes creatively reformulate given conditions to produce novel and inventive solutions."[39] Though datascapes may resemble actual architectural or urban design projects, one must be aware that they are not actual designs, although they may come close to them. Brett Steele claims that datascapes "as the name suggests might offer opportunities for escape . . . and they ultimately offer the possibility of dealing with space."[40]

MVRDV's *Metacity/Datatown* is a research and artistic project that wants to "explain to a large audience what the spatial consequences are of certain policies that deal with food production and waste, among others."[41] In a more recent approach, MVRDV took the ideas of *Metacity/Datatown,* and examined space creation based on increasing capacity for people and consumption. With an approach based on experimentation and extrapolation of data, MVRDV uses an interactive "planning" device (the computer) in which statistical and territorial data are used to generate simplified spaces. Calling this process 3D City (a datascape of the evolutionary city), MVRDV claims that in using this process "everyone is a citymaker"[42]:

> The idea of developing user-interactive "planning machinery" becomes more attractive and more necessary. Statistical and territorial data are currently filed on the web. Analysis and monitoring systems are progressing . . . (software packages) would enable them [planners, development agencies, community centers, and political parties] to find data as well as to communicate, control, discuss, debate, evaluate, and protest. . . . this device can select, sort and combine data and illustrate processes.[43]

Maas has found an alternative means to introduce information as part of an urban design. Datascapes present a new vocabulary for urban design. They establish a semiotic connection to information through form. The manifestation of MVRDV's work is highly abstracted and critical for architectural discourse. Maas's datascaping approach is interpretive and also a means for describing hypothetical design experiments.

In contrast, George Weinberg, a guest journalist for the *Yale Herald* comments on MVRDV's datascapes:

> The works are interesting and thought provoking, but MVRDV's treatment of them as objective research is misleading, and counterproductive.[44]

Weinberg goes on to say that "most of the works appear more like utopian visions than self-described objective research."[45] In other words, Weinberg believes that the calculations and statistics presented by MVRDV are not scientific. Can this method really be used as an analytical tool? Weinberg is concerned that Maas's datascaping projects depict unrealistic conditions or far-fetched scenarios that just become a crafted vision from Maas's imagination and serve no purpose other than as architectural rhetoric. They may seem unreal – that is perhaps what makes them surreal.

MVRDV's *Metacity/Datatown* and other subsequent datascape projects (Pig City, 3-D City) do bring interesting data to the forefront in a provocative way. The datascapes let us think more intuitively about our living space. Though some of the data is so far-fetched, the experiment is taken to a level of creative mapping that often dreary graphics from other standard data visualization tools do not provide. Maas argues: "I think it's important to show that there are more ways to think about (data) and solve problems (visually)."[46] These individuals have made significant contributions to the role of mapping as it relates to art, urban design and information display, through new modes of computer-generated representations. In fact, they have bridged the gap between these disciplines in a harmonious manner, which has produced powerful digital maps as an end result. Together, these individuals form an international roster of architects and artists using digital means to create maps that define spatial parameters. Their works celebrate the dynamic relationship between urban context, space and data organizations.

Dramatizing the data

Finally, it is important to consider that MVRDV's method could be applied to a number of urban data-driven situations, taking note of their styles of representation. When creating datascapes, one important factor to consider is the extent of distortion. Distortion is part of the artistic license, which if handled properly helps to clarify the information that is communicated. By creating vertically extruded peaks in *Datatown* for example, the volumes of these cones communicate high volumes of waste, water consumption or other factors. This process is conceived in a way that still carries the message of the datascape in an informative manner, but which does leave room for interpretation. Datascapes created by MVRDV are precedents for other architects and students to investigate, review and further develop, using more advanced techniques and alternative artistic touches. Through powerful means of visual communication datascapes assist the viewer or city dweller in understanding his or her city and its possible urban situations. The next chapter focuses on the role of creative modes of mapping the invisibles, in particular the works of James Corner, through an artistic and informative approach. He takes a poetic, qualitative, even deceiving approach to mapping, in order to draw attention to features that might otherwise be missed in the landscape.

An interview with Winy Maas: some final thoughts[47]

Nadia Amoroso (NA): How did you first derive the concept of the datascape?
Winy Maas (WM): I came up with the concept about fifteen years ago and it stemmed from a couple of components that are interesting to mention. First, it was at a moment in which architecture and urbanism was based on French philosophies about complexities, and observing and studying these

complexities. Datascapes give a more mathematical answer towards the complexity. The second component, which was surrounding *meta* (data types and kinds), deals with larger urban processes that became more relevant in the practice again. Architects, like me, were only dealing with objects. There was a need to redefine architectural object through some meaning in an urban and larger format, and urbanism deals with issues of numbers and statistics and of course larger scales, more than the individual. The demand for a numerical approach or an interpretation was needed for a larger scale in which objects can be situated within this. The third component is about the urban realm and mapping, and how to connect these with urbanistic studies. . . . Many institutions were doing this like the AA and Princeton. I wanted to see what we can do with this, pushing the mapping approach. Usually mapping is a confirmation of the existing and we needed a next step with this. . . . for me, these are the basic concepts of the birth of the datascape.

NA: In your datascape representations, specifically those in *Datatown*, to some extent the data are dramatized but nonetheless still hold true to the numerical facts. Do you think this is needed to carry across the message of the space they try to represent?

WM: Yes, dramatization can help to bring out certain urban urgencies or necessities. Therefore, yes, you are completely correct . . . without trying to falsify too much, or at all, so there is a need for an apocalyptic approach to get the urban environment moving. We can easily drive more cars, and when we are completely jammed, we can maybe do something about it. For climatic reasons, no problem, we can drive electric cars or hybrid cars. So it is only a problem of capacity at the moment. Therefore, apocalyptic suggestions or aspirations can help to discuss what should be done, what is valid and what is useful. Secondly, in dramatization, this is done to investigate thresholds, maximizations, and later optimizations. . . . to see where and at what moment you can do certain things, and to visualize this. I think that is the important moment, beyond the mapping, which is rather neutral and is only about place, and now [the mapping] becomes three-dimensional and the capacity issue becomes much more "aware." For example, we worked on a project for UNESCO . . . to show certain thresholds or X number of densification envelopes using UNESCO laws or by-laws. This [mapping and dramatization] helped visualize the envelope and space. Through this densification envelope, one was able to see if this was enough density or not, based on the law. Through the datascapes, one could see if the law needed to be changed or if it was fine as it was currently. This was an investigation about limitations. . . . to see these limitations and see if one can go beyond this point. It is about showing the laws and the by-laws. These rules are fine, I have no problem with by-laws, and they are basically a tool for collective space or collective building practices or urbanistic practices. One can say that the rules or by-laws are actually notations of spatial culture. It shows what kind of laws or rules society wants to have. Mostly those rules are good, but up to the moment that they may limit our potential or next steps. Therefore, maximization helps to "reveal" the threshold between the moments of acceptance of these rules or the applications of other rules, and to accommodate our society and the future city. Therefore, dramatization is found at points of maximizations.

NA: Do you think your "datascapes" are some kind of three-dimensional maps or guides into the invisible city? In some respects, is there some connection between your datascapes and Hugh Ferriss's New York zoning Ordinance drawings of the 1920s in revealing the envelope of this by-law? (Of course, you are using digital means and contemporizing this approach.)

WM: Yes, I can see this connection with the statistical reference and I see the connection with his work. . . . Due to better computations within both drawing and numerical environments, we are able to accelerate these processes and be more precise. We are able to handle more data in that way. Yes, it [the datascape] is the next steps or it is a "contemporization" of that epoch, and now this [datascaping] is for this era. There is now an enormous and wide potential. . . . We don't export some of them [data], so there is not that complete realm. Therefore, the next steps after datascapes are to put all the data in a coherent way. So, what we did with our datascapes in those days was spatial. The zoning datascapes we did for UNESCO which did look a bit like the New York ones [Ferriss's drawings], though, his [Ferriss's] drawings were more Gothic looking. What we were doing [for UNESCO] was showing three-dimensionally what to expect in European cities and their future. The envelopes we did for UNESCO were very three-dimensional. We also showed a design proposal for a new suburban neighborhood based on individual parcels of land. In *FARMAX*, we used statistical demands to encourage diversity and to notate this and to show the outcome at the end. We used six or seven parameters and faculties [data input]. When we "spatialized" these seven or so faculties, they easily led to hundreds of faculties and these were translated into the design for the suburb project. This was based on theories and mathematical sequences. We were trying to make datascapes better. From this, we derived optimization tools which helped monitor and generate datascapes like in the *OptiMixer*, *Climatizer* and later *Regionmaker*. It used a pixilation method which showed the data in three dimensions. This was much more developed and can absorb all spatial rules and by-laws. I found this as an interesting iteration or language, where all these datascapes can be generated, and turned into one connective system and method. We are still working on refining this process.

NA: What do you think are the future applications of datascapes in architectural and urban designs?

WM: If I follow the line of the application of this software, it is one of the tools that can be extended endlessly, I guess, and I think MIT does a good job in these kinds of tools, in technology and to make them evolutionary. Those kinds of processes are on their way from a theoretical point. . . . I love that. This can lead to an enormous potential in the software, which could really be used in design processes. For climatic purposes, you can see when there are calculations on temperatures in certain areas. We can see how in structural engineering, optimization methods are embedded, and these modeling methods are used in other professions. If we accelerate the initialization processes then this becomes easily "architecture and urban design". . . . if we automate this process, using some kind of intelligent software package, like many other professions use, we can visualize these kinds of information more easily in the world of architecture and planning. I think that's one of these things that we need more efforts in. So, it is one of the things we need to do next . . . and then this can lead to fantastic programs like datascapers that can be applied to architecture and urban design. I believe IKEA does this with their interior home design software . . . you can make your own house designs. Therefore, this is one area which I think it is in and can be developed. On the other hand, in terms of architecture, say in the cultural dimension, how explicit should architecture be? We can notate a "zone" like what we [MVRDV] and what later on the Dutch movement was doing . . . then there were a lot of people interested in this idea, and were working on this kind of "communicative architecture." More than before, they wanted to express a kind of "directiveness" and show the frustrations or the limitations of the faults of laws and rules. There is still an area to be developed enormously. At that moment architecture becomes more "directive" and datascapes are a device that can help this and make that moment in architecture become a "communicative part." This is one of the ways to give architecture, again, a better position in society and make it more valid. So, this is one aspect to explore. The good thing about this trajectory is that any object can be simple and "reductive." It is not as complex as we first pointed in the software. This complexity turned into a "reductive" aspect will help bring it [the datascape] to new summaries, new certifications, and can lead to new moments that are more communicable and therefore emblematic . . . I think this is a highly interesting aspect to analyze and to develop.

Notes

1 Brett Steele, "Reality Bytes: Datascapes" in *Daidalos* 9/70, 1998, p. 10.
2 Bart Lootmsa expresses his thoughts on the concept of datascapes in *Berlage Cahiers* 7, July 1997, p. 39.
3 James Corner, "Operational Eidetics: Forging New Landscapes" in *Harvard Design Magazine*. Fall, 1998, p. 25.
4 Ibid.
5 Bart Lootsma, "The Diagram Debate, or the Schizoid Architect" in *Datascaping*, http://predmet.arh.unilj.si/siwinds/s2/u4/su4/S2_U4_SU4_P6.htm (accessed June 2007). Abstract systems are another topic of discussion that I will touch on briefly; I am more concerned

with the visual aspects of the datascapes, though their political impact in society is important.
6. Bart Lootsma, "What is (really) to be Done?" in *Reading MVRDV*, edited by Veronique Patteeuw, Rotterdam: NAi Publishers, 2003, p. 29.
7. Ibid., p. 8.
8. Ibid.
9. Ibid.
10. Curtis D. Frye, *Metacity/Datatown* book review, 2000. http://www.techsoc.com/metacity.htm.
11. Taken from the back cover of the book *Metacity/Datatown* by Winy Maas (Rotterdam: MVRDV/010 Publishers, 1999). *Metacity/Datatown* is based on a video installation of the same title produced by MVRDV for the Stroom Center for the Visual Arts in 1999. The publication was created afterwards. The project was conceived by Maas, and researched and produced by the firm MVRDV, including Jacob van Rijs and Nathalie de Vries.
12. Maas, *Metacity/Datatown*, p. 19.
13. Ibid.
14. Ibid.
15. Ibid.
16. The edge length of *Datatown* is defined by the equivalent of someone traveling for one hour by bullet-train (400 km/hour). *Datatown* is claimed to be the densest place on earth, with over 240 million people. It is comparable to the USA contained in one city.
17. Ibid., p.18.
18. Ibid., 39. Extreme scenarios are "what–if" situations using extremely high levels for the six data sectors.
19. Ibid., p. 65. Maas calculated a total volume of close to 44 billion cubic meters allocated for the "Living Sector."
20. Sven Lutticken, "MVRDV: Stroom HCBK – The Hague – architectural firm exhibition" in *Art Forum*, December 2001, http://findarticles.com/p/articles/mi_m0268/is_/ai_80856220.
21. Ibid.
22. Ibid.
23. Immanual Kant, "Book II: Analytic of the Sublime – The Mathematically Sublime" in *The Critique of Judgement*. Oxford: The Clarendon Press, original print 1790, reprint 1964, section 25, p. 94.
24. Ibid.
25. Weller makes another important point about the *Datatown* project. Maas does not extrapolate current existing statistical conditions and "follow function with form." Though Maas uses statistics to examine the restrictive spatial condition of the hypothetical town, the results of his experimental project are at the opposite extreme from McHarg's "ideal of a static culture" finding its fit within the landscape's limits. See "An Art of Instrumentality: Thinking Through Landscape Urbanism" in *The Landscape Reader*, edited by Charles Waldheim, p. 82.
26. Theodore Levitt, "The Globalization of Markets" in *Harvard Business Review*, May/June 1983, p. 99.
27. This discussion was a result of a symposium among a group of artists, architects and planners, including Winy Maas, on the topic of Numbers and Culture. Visual artist Ronald van Tienhoven advises the Mondrian Foundation on issues concerning art in public space. He consulted Maas on the installation of the *Metacity/Datatown* project.
28. Maas, *Metacity/Datatown*, p. 214. Van Tienhoven describes the need to incorporate statistical information concerned with the city into art and architecture.
29. Winy Maas and Jacob van Rijs, "Datascapes 2" in *Berlage Cahiers* 7, 1998, p. 38. The hidden logic behind urban matter can be defined as the invisibles of the city – elements that cannot be seen yet hold urban information. Maas and van Rijs present their students with the challenge to define and analyze datascapes, and to produce extreme conditions that

generate new city forms shaped by the given data parameters, much like Maas's explorations in *Metacity*.

30 Ibid.
31 Lootsma, "Synthetic Regionalization: The Dutch Landscape Toward a Second Modernity" in *Recovering Landscape: Essays in Contemporary Landscape Architecture*, edited by Corner, p. 257.
32 Ibid.
33 Weller, p. 83.
34 Ibid.
35 Ibid., p. 83.
36 Alicia Imperiale, "Refiguring the Figure: Imaging the Body in Contemporary Art and Digital Media" in *Mapping in the Age of Digital Media: The Yale Symposium*, edited by Mike Silver and Diana Balmori. West Sussex, UK: Wiley-Academy, 2003, p. 35.
37 James Corner, "Operational Eidetics: Forging New Landscapes" in *Harvard Design Magazine*, Fall 1998, p. 25.
38 Ibid.
39 Ibid., p. 26.
40 Bart Lootsma, "Towards a Reflexive Architecture" in *El Croquis: MVRDV*, vol. 86, 1997, p. 39.
41 Bart Lootsma, "*Biomorphic Intelligence and Landscape Urbanism*" in *Topos*, 40, 2002, p. 16. Lootsma takes MVRDV's datascape of Pig City as an example of this argument. Pig City reveals the spatial consequences of organic pig farming, and proposes alternative solutions that take shape through self-sustainable high-rises. Through this form, MVRDV's images demonstrate a mix of uses of food production for the pigs, and slaughterhouse.
42 MVRDV, "Everyone is a Citymaker: Optimizations" in *MVRDV: KM3: Excursions on Capacities*, pp. 1252–53.This research publication explores issues in the development of 3DCity. It investigates attitudes and observations of increases in global densities, via hypotheses for how to deal with them, and the creation of vertical spaces, through speculations on "localizing global densifications." The book also examines emergent spatial possibilities (based on hypothetical dense cities), via experimental 3-D mapping and project-based scenarios.
43 Ibid.
44 George Weinberg, "Dutch Firm Envisions City of the Future" in *The Yale Herald*, 6 September 2002: 2.
45 Ibid., p. 1.
46 Iovine, "Dutch Designs for Cities Built on Ideas and What-Ifs" in *The New York Times*, 15 October 15, 2002: 1.
47 An interview with Winy Maas was conducted in August 2009, discussing topics relating to his datascapes, spatial mapping, their visualizations and their future applications.

4 The Map-Art
The Works of James Corner

This chapter discusses the process of landscape mapping and offers some visual interpretations of such maps. In particular, it profiles the peculiar mapping process and art created by James Corner (landscape architect, professor and contemporary landscape theorist) and its role in landscape design.

It is necessary to first talk about the late Ian McHarg, the well-known American landscape architect and environmentalist. As a graduate and now director of the University of Pennsylvania's Landscape Architecture Department, Corner was influenced by the works of McHarg, Director of the School in the 1960s. McHarg is known today for his methods of capturing those natural and social aspects of the sites that drew his attention and concern. These aspects included vegetation, hydrology, soil structure, geology, morphology, sun and shade areas, erosion, areas of sensitivity, and other natural components related to the site. His sites were often large in scale – a necessity, in his view, if one was to understand the entirety of the landscape process. He performed complex site analyses of these physical attributes. He separated areas of the site that displayed all the above aspects into individual sheets (sometimes transparent sheets) and then layered these individual maps. Through this layering process, an overall master map was created that allowed further understanding of the site. McHarg used this scientific, analytical and drawing process to "capture" the site before he attempted to design for it. This method of drawing to reveal site issues was a breakthrough technique in landscape architecture and became a new mode of representation to reveal specific data pertaining to the site. The data usually applied to physical conditions, which were normally difficult to make accessible and comprehensible. Nonetheless, McHarg made this information readily available and easily comprehensible through color coding. This type of visual communication allowed the viewer to become familiar with the site. Today, his overlay system is known as the McHargian mapping system and it is still a vigorous and vital part of the landscape mapping field.

Although times have certainly changed, and although the techniques have since advanced to make use of technologies like GIS (Geographic Information System) and other digital mapping processes, the chief desire – to illuminate natural and social information that other forms of mapping might overlook – has not changed since McHarg's groundbreaking work. Additionally, the mode of representation found in his work, while undeniably impressive, nonetheless lacks poetic flare to the modern eye. The mode of representation was simple, usually using a hatch pattern or colored marker to represent a landscape condition; for example, a dark brown might equate to steeper slopes whereas a lighter shade might represent gentler slopes. However simple the style of representation might seem to us now, McHarg's site-analysis mapping concept endures today because it showed the potentialities to convey site information in layered categories and color coding.

McHarg's lineal descendant in the field of urban mapping was James Corner, and Corner's work warrants investigation in the field of mapping. According to Corner, mapping is a process that involves a "complex architecture of signs"; it is, in other words, a "visual architecture" through which the space represented is selected, translated, organized and shaped. Throughout the experimental phase, the proposed outcome constitutes a rethinking of maps by introducing an artistic component to the work, and presenting a closer tie to urban design. James Corner's great contribution was that he took the foundation laid down by McHarg and added to it – substantially.

Decades after McHarg, James Corner reinvented landscape mapping styles so that they might achieve more poetic and seductive levels of representation. In 1996, he and famed photographer Alex MacLean produced a compelling publication, *Taking Measures Across the American Landscape*. This book profiled various landscapes throughout America with the use of aerial photography. Relic landscapes, agricultural lands, railway sites, mountainous terrains, wind farms and burning fields (among other subjects) were captured through MacLean's artistic use of the lens. Delicately composed, these images tell the story of each place. The photos are captivating images but, accompanied by Corner's supplementary "map-drawings" (a combination of map and drawing styles such as collage crafted by Corner as a means to interpret the aerial photos, and thus, creating a map-drawing of the site), the aerial photos become instructive as well as aesthetically pleasing. Corner's drawings and photos provide expressive and informative visual data about the sites he profiles, through his alternative and artistic approach to mapping. Through his creative process, Corner, as the mapmaker, selects which aspects of the site are revealed and what is left to the imagination.

It is instructive to bear in mind Tufte's concerns about graphical integrity and why so much distortion and deceit exists in maps[1] – distortion that invariably diminishes the credibility of the graphic data and complicates efforts to arrive at some greater understanding. In Tufte's view, "Graphical elegance is often found in the simplicity of the design and the complexity of data."[2] Tufte is also of the mind that the mapmaker should select a proper format and design in portraying complex data in an aesthetic fashion. And he is firmly wedded to the notion that it is advisable to eschew superfluous content – mostly, free decoration and "chartjunk." With this in mind, how does Corner portray complex landscape/urban information in his maps?

The map-drawing hybrid

While not without flaws, Corner's works have set a great precedent in contemporary landscape architectural discourse as it relates to mapping and visual communications. Corner, in partnership with MacLean, has created non-traditional diagrams that map the site – "non-traditional" meaning that they are not the two-dimensional standard line drawings or sketches that make up bubble diagrams, with stars or asterisks for nodes – much as in Lynch's mapping of the city. Rather, Corner's map-drawings are beautifully drawn or digitally altered diagrams of various "fragments" or contour lines in the landscape, that are reassembled almost like a collage to tell a story; consequently, the images are both informative and artistic in quality.

To start with, in "Survey Landscape Accrued" (14" × 20"), Corner embeds MacLean's aerial photograph within a corner of the picture (Figures 4.1 and 4.2). This is a rare technique in Corner's map-drawings, but one that has eye-catching benefits. First of all, with the insertion of this photo, the audience is immediately able to understand that the interpretive piece (the drawing) is based on the embedded photo. The image depicts patterns of settlement, parcel size variation, geometry of the land subdivision and plant variation. The simple white line grid and square patterning obviously reflect the landscape geometry and its various parcel sizes. In many of Corner's map-drawings, contour lines are superimposed on pictures of the landscape; in fact, it may be said that this is Corner's signature style. It seems as though landform is a landscape feature that he believes (as a landscape architect) is noteworthy of conveying in his map-drawings. Further, while not entirely clear, it may be inferred that the small collaged pieces of flowers and other plant material relate to the type of species that grow within the parcels of land. In another interesting aesthetic flourish, a ribbon strings its way across the piece, indicating the river seen in the photo. All in all, the drawing is beautifully rendered – though Corner does not use thorough labeling in this map-drawing (and in many of his drawings) and this makes the data presented somewhat elusive. This falls short of Tufte's injunction that graphical integrity demands the labeling of data where needed.[3] Corner's map-drawings are artistic in character, but fall some way short in the informative aspect of map-making. However when accompanied by MacLean's aerial photographs, the *map-images* are clearer to comprehend. But even with this wholly acknowledged, one must still respect the stance taken by Michael van Valkenburg:

> The collision and layering of the pieces of today's landscape, like fragments of our perplexing and complex social self, are yielding a

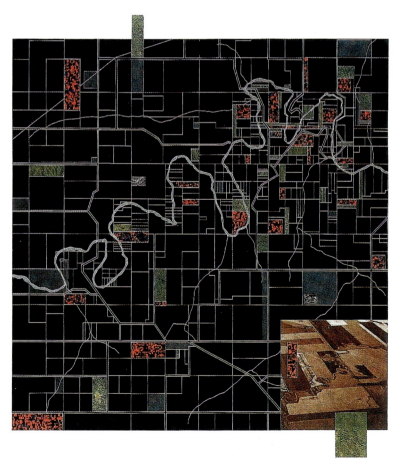

4.1 and 4.2
"Survey Landscape Accrued," North Dakota, USA. Drawing by James Corner and photograph by Alex MacLean.

Source: James Corner and Alex MacLean, *Taking Measures Across the American Landscape*, Yale University Press, 1996, pp. 50, 40 (respectively)

remarkable mosaic, one that speaks with clarity and power to interwoven and complex relationships between our culture and the land. Alex MacLean's photographs and James Corner's collage overlays and writings reveal some of the origins and processes of the patterns that are still evolving in the contemporary American built landscape. Their remarkably vivid images provide clues for deciphering the sources and qualities of the constructed landscape.[4]

The advanced diagrams that accompany the striking aerial views provide the viewer with an alternative way of interpreting the sites. Artistic in quality, the map-drawings allow the reader/observer to better understand the land they are observing through deeper insights, but still leave him or her with unresolved questions about the site. These sophisticated diagrams are meant to be viewed as metaphorical and speculative works that are intended to provoke thought about the potentials of the American landscape.[5]

What is interesting about these visual recordings is the combination of the drawing and MacLean's stunning aerial photography. As much as anything, these drawings are intuitive figures that suggest deeper insights into the location than were possible previously. Also there is a particular immediacy to the photography and the drawings that allows the viewer to be drawn into the site. With a map there is invariably some level of measure or precision of scale to represent the site. Corner's map-drawings or sophisticated diagrams, however, successfully break free from the tyranny of conventional maps; in a real sense, the maps yield what might be described as "imaginative forms of measure."[6] In the final analysis, the combination of the map-drawing and the aerial photograph is a powerful tool for investigating the hidden elements of the site. Furthermore, to those who might feel that Corner's work is too symbolic and/or abstract, his response is quick and trenchant:

> Consider the role of maps, for example, which, like aerial photographs or paintings, are documents that are not remotely like the land itself; they are flat, uni-dimensional, and densely coded with all sorts of signs and hieroglyphs. To read a map, one must be trained in cartographic conventions. . . . Maps make visible what is otherwise invisible.[7]

Those who express unhappiness with the artistic elements of Corner's work seem to miss the essence of what map-making is all about; the incompleteness of all maps which necessarily demands some intuition on the part of the viewer since only certain information about the site can be provided in a representation that is fundamentally uni-dimensional and essentially hieroglyphic in nature. In a sense, art is an inescapable part of map-making and may be the only means by which anything approaching full disclosure of a site can be achieved. Corner carefully masks certain realities of the site and reveals only what he (the mapmaker) wishes to expose.

One good example of such artistic license can be seen in Figure 4.3, "Burnings" (12" × 16"), in which Corner paints a section of burning field between two major contour lines of the land; as he does this, Corner also indicates winds from the east-south-east. There is assuredly an element of "truth" in all this, but the work is, first and foremost, an artistic composition that happens to have some factual elements added in. Possibly the most "artistic" component of the illustration is the dark brown smear painted across the base-map: this is fundamentally intended to depict the potency of the burn across the site. Naturally, this sort of artistic flourish underscores an important truth that Corner wishes to convey (which is why art can well serve landscape mapping). However, Tufte's principle of clear and detailed labeling is not apparent in most of Corner's map-drawings – although he does label several major elevation points on the drawing, indicating the location and dimension of the burn. Corner does not use thorough and detailed labeling within his map-drawing because such text will detract from the pictorial quality of the map. Corner's maps have

4.3
"Burnings." Drawing by James Corner.

Source: James Corner, *Taking Measures Across the American Landscape*, Yale University Press, 1996, p. 109

metaphorical, symbolic and spatial powers that visually reinterpret the elements or sites they represent even as they underscore and/or emphasize truth.

Creativity in mapping

Corner describes the process of mapping as an "agency." An agency, by definition, is an operation, condition, or state of acting or of exerting power. An agent is the vehicle through which power is exerted or an end is achieved. The agent becomes the instrument of achieving something. The agency becomes the office or function of an agent: the relationship between a principal and his or her agent.[8] The agent is a powerful mechanism that pushes for the end result. According to Corner, maps are graphical registers that depict points and places in space; they take the "measurement of the world." They perform a specific task, revealing the unknown:

> mappings are constructed from a set of internal instruments, codes, techniques and conventions, and the worlds they describe and project derived only from those aspects of reality that are susceptible to these techniques, are dimensions of mapping still barely understood by the contemporary planner. Instead, most designers and planners consider mapping a rather unimaginative, analytical practice.[9]

Mapping should encompass creative power, knowledge (information) and inventiveness. In particular, Corner is interested in "mapping as a productive instrument, a world-enriching agent, especially in the design and planning arts."[10] He believes that the mapping process is far more powerful and knowledgeable than its tracing agency. Tracings (or maps that trace) simply reproduce what is already identified. These maps define information visually but expose nothing new. Part of Corner's concern is how "new things" can be represented in maps. One particularly important question raised by Corner – and it is one that should be uppermost in the minds of all those concerned with making landscape mapping as insightful as it can be – is the following: As new site information or site elements unfold, do a set of new eidetic images emerge to reflect these elements? Graphical inventiveness and innovation are just as important as the type of "obvious" urban information these representations expose. Corner's maps stand as both images (constructed paintings) and as sets of instruments that exhibit critical data. While not in themselves definitive urban designs or landscapes, these maps do resemble such designs. With a dash of creativity, architects can play a key role in the final detailing of an urban location.

Maps are graphical representations that help us understand conditions, processes, or events in the human world. Maps are visual representations of the real world or a part of it at a reduced scale or magnitude. They are representations of places, spaces or *conditions*. Because of the extreme change in size between reality and its mapped geography, the mapmaker is selective of what to show. The map cannot represent all of "reality" in absolute terms, only parts of it.[11] Therefore, the mapmaker is selective of what to show and how to show it, sometimes making maps subjective. They are often considered visual instruments that guide the individual through the parts of the geo-reference they depict. Must a map be two-dimensional? Not necessarily, but often a map is depicted on a two-dimensional surface such as a sheet of paper. Maps often represent three-dimensional reality in a two-dimensional format. Maps to some extent must have some form of mathematical projection. "The mapmaker employs various cartographic devices, especially 'scale' and 'projection.' Most maps have formal elements printed right on the map that give you guidance about how the mapmaker has represented the scene: directional information, keys, and scales are part of most maps."[12]

Maps can be thematic, that is, showing social, political, cultural, economic conditions that can be geographically based, or not. A map can be analytical, allowing individuals to help draw conclusions based on the map's representation. Sometimes, a "single map is not very effective at showing 'process.' A narrative can move a reader through time very quickly; a map tends to be static and to show a single place at a single moment. A map then, in distinction from written texts, can be understood as privileging place over process, contextuality over linearity."[13] Perhaps animated maps can help solve this issue. A map in some sense is trying to convince the reader of some particular aspect the map is attempting to represent. In one sense, the mapmaker is making an argument about a particular aspect he or she is trying to represent. He or she may use a particular style or graphic device to communicate these aspects.

According to geographer Gerald Danzer:

> You can start looking at what's on the map, what is not on the map. What is missing? How is it packaged? How is it framed? How is it meant to be used? We don't know a lot of these things, by the way. It's like any historical document that you get. That it tends to all of a sudden put things together spatially, and say, "Well, what implication does it have? What does that mean? Why is this cartographer emphasizing this dimension?" . . . And we might, in our minds, think about alternative ways, alternative audiences, alternative purposes. And once you start playing with the map like that, it's a series of questions and answers and arguments, and delivery systems if you wish.[14]

Maps are graphical devices that reveal evidence of particular parts of reality or particular conditions. They can be packed with information or a selection of information. The mapmaker has the choice to select what needs to be revealed or not. "Maps present . . . an eidetic fiction constructed from factual observation. As both analogue and abstraction, then the surface of the map functions like an operating table, a staging ground or a theatre of operations upon which the mapper collects, combines, marks, masks, relates and generally explores."[15] Corner is stating that maps are abstracted images synthesized from factual material (i.e. geographical reality);

however, the mapmaker selects and chooses what to display in his or her map and how to display this information graphically.

Can aerial photographs be considered maps – and to what extent? This is another important question that leads us directly into a discussion of Alex MacLean. MacLean combines factual observations into what Corner claims "an eidetic fiction." MacLean carefully frames certain parts of the landscape that he wants to represent. Architect and theorist Julia Czerniak argues that in the works composed by Corner and MacLean, by adding aerial photography, the depiction of landscape becomes more than static images. The aerial photography reveals how landscape "works as a process" and as a "continuing activity and set of relations that change over time."[16] Czerniak goes on to argue that the addition of aerial photography to maps of landscapes (as in the Corner and MacLean work) depicts conditions of the "invisible from the ground" and presents landscape as a "complex network of material activity" rather than a "static and contemplative phenomenon."[17] Corner defines his map-drawings as a composite of "maps and photographic images overlaid with invisible lines of measure such as logistical equations that explore how 'landscape representation (particularly that of aerial view) not only reflects a given reality but also conditions a way of seeing and acting in the world.'"[18] Corner applies "practical and poetic" measures in his map-drawings. Practical measures relate to the scale and mathematical projections of the landscape, while poetic measures relate to artistic interpretations, and can also relate to cultural aspects of the landscape. Czerniak explains that "the aerial photograph makes apparent existing landscape elements such as the persistent geometry of the grid, while the map-drawing renders visible what is latent: planning abstractions, scientific data, and poetic pre-occupations."[19]

The aerial photo and the map-drawing

In the case of Corner and MacLean's work, aerial photography is an auxiliary visual instrument for the reader/viewer in order to assist him or her about the landscape. It is not a "true map" per se in the sense of having measurement, projection and scale, but coupled together with Corner's map-drawing, the photograph is more informative. On its own, MacLean's aerial photographs are works of art. Landscape critic and geographer John Jakle comments on MacLean's photographs (from *Taking Measures*):

> The good photograph conditions what we see, and how we see it. The aerial photo, for example, distances the viewer and, as such, miniaturizes the world with map-like fixation. But these photos also stand as works of art, drawing and tantalizing the eye through effective composition, use of color and sense of framing.[20]

As described in Corner's book,

these map-notation drawings are meant to complement the photographs and at the same time to stand apart from them. Both types of images are aerial, offering a synoptic perspective, but the map-drawings play on certain planning abstractions, such as making visible strategic organizations of elements across a ground plane or revealing certain scale and interrelational structures (from regional to local networks of communication, for example). Consequently, the drawings (but also the photographs, in a different and less explicit way) reveal how typically prosaic and analytical methods of synoptic planning and land systematization harbour a more poetic, creative potential."[21]

Corner's interpretive map-drawing (Figure 4.4) entitled "Windmill Topography" (14" × 20"), carefully conveys the beauty of the California windmill ridge line, captured in MacLean's photo (Figure 4.5). At the left hand corner, Corner applies a blue wash strip that profiles the form of the ridge line. Superimposed on this strip, Corner creates a series of little white pin-like windmills that stand out sharply against the darker background. He lightly draws an enlarged profile of the windmill blade, depicting the element in detail. Immediately, one is taken by the fact that the map-drawing is filled with information for one to decode. Thin red lines marked across the sheet show various wind directions and wind pressures. To the right of this image, two modern wind turbines stand proudly in the foreground of the page. They are anchored onto a narrow strip photo of the site; the textual information is also superimposed onto this strip. Moreover, a horizontal black strip (it is found three-quarters of the way down the page) lines the picture perpendicular to the large windmills on the right-hand side of the illustration; this black strip is a section of the ridge line that successfully captures its mountainous terrain. Embedded in this black strip is a minute army of windmills that highlight the ridge-like form of the site. The letters "LA" are branded into the left hand side of the piece, telling the viewer that Los Angeles is located towards this side. Behind the black sectional profile of the ridge line, and completing the tableau, Corner chooses to place a wind pressure chart – again emphasizing the role of wind in shaping the features and potentialities of this location. The piece as a whole is informative and attractive; more than that, it is rich with information for the viewer to decipher – information that may entice the viewer to learn more about the site and about its recurrent phenomena. Once more, it cannot be overstated how Corner uses little or no labeling with his map-drawing. Labeling, much to the chagrin of Tufte, will only hinder the artistic intent of the drawing. Instead, Corner relies simply on the supporting caption below the landscape photograph or the map-drawing for further clarification.

Corner often takes a poetic, yet deceptive mapping approach to bring across elements of the landscape that otherwise would be left unnoticed. This approach is important in the field of landscape architecture in order to visually deliver the critical aspects of the landscape – a guide to hidden landscape and a resource to pre-design concepts. In the map-drawing of northern plains of Montana, entitled "Dry-Farming

4.4
"Windmill Topography."
Drawing by James Corner.

Source: James Corner, *Taking Measures Across the American Landscape*, Yale University Press, 1996, p. 83

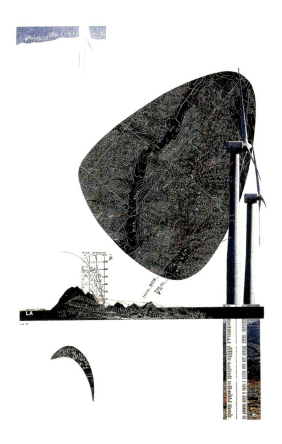

4.5
"Windmills along Ridge Lines, California, USA." Photograph by Alex MacLean.

Source: James Corner, *Taking Measures Across the American Landscape*, Yale University Press, 1996, p. 82

4.6
"Dry-Farming Wheat Strips," Montana, USA. Drawing by James Corner.

Source: James Corner, *Taking Measures Across the American Landscape*, Yale University Press, 1996, p. 125

Wheat Strips," Corner carefully captures the endless horizontality of the field strips (Figure 4.6). The image captures the notion of this productive landscape by the positioning of the carefully maintained vertical strips of wheat that dominate the image. The harvester positioned in the top left-hand corner says something about the overall maintenance and production of this farmed landscape. Corner has his trademark white ink on black contour plan positioned in the central part of the picture. Superimposed onto this map are two prominent collaged strips that represent the long and narrow wheat strips running north–south along the plains. In a sense, despite the asymmetry of the composition, the pervasiveness of vertical lines gives the composition a sense of balance and proportion that is not dissimilar to some qualities one finds in MacLean's aerial photos. In partial explanation for his use of the strips, Corner notes, "In the Northern Plains, strips of wheat run north and south. . . . sometimes a mile long and only 140 feet wide."[22] It should be added that, in one of the square mile grids of the topographic plan, two smaller-scale strips depict the mile-long strips. On the bottom right-hand section of the picture plane a series of green strips stands vertically, casting a forward shadow that communicates the dominance of these elements in the landscape. To the top left-hand corner of the canvas, a small image of a harvester tells the viewer that this machine is required to create these precise strips. As Corner puts it, "For many artists, it is the inventive capacity of representation that enables them to provoke new and alternative ways of seeing the world."[23] Critic and geographer John Jakle explains,

> James Corner's diagrams do attempt to lead the reader/viewer into and out of various photographs. They are especially effective in this regard when used to illustrate the national land survey. They were drawn to be displayed at a much larger size, however, and their reduction for the book can only be partially remedied by the judicious use of a magnifying glass.[24]

In essence, the map process becomes a quest to invent creative means of depicting elements that would otherwise have escaped notice. Corner's map-drawings become both the subject and object of representation; in and of themselves, they are carefully crafted pieces of work – even paintings. They can become objects that stand in a gallery, waiting for the viewer to see them and admire them on strictly aesthetic grounds; their subjects are the elements of the land they try to represent. Corner writes that, "Mapping is a fantastic cultural project, creating and building the world as much as measuring and describing it."[25] One is left with the impression that Corner is aware – to an extent most are not – that cartography or landscape mapping are really artwork, and that art is what allows people to develop a multi-layered and holistic understanding of the world around them.

According to Corner, one of the main properties of mapping is that of agency – it is an agent of speculation and critique. He believes mapping has the ability to emancipate possibilities, to "enrich experiences and diversify worlds."[26] The process of mapping is a powerful act, one that unveils hidden potentials, an act that reveals knowledge of the field conditions. From his perspective, Corner sees mapping as "a productive and liberating instrument, a world-enriching agent, especially in the design and planning arts."[27] The power of mapping can be invigorating and exciting. It has the power to evoke our curiosity. It allows us to become curious about the unseen and it instills in us the wish to unveil more and more about the unknowns of the hidden city. More than all that, mapping contains dual characteristics – one is the analog representation of ground conditions and the second is the abstraction of these conditions (codification, selection, projection, etc.).[28] This dual function of mapping presents quantitative and qualitative "markings" of the site. In particular, the surface map can hold information on socio-economic factors, ecological data, cultural trends, and global patterns, while at the same time painting an aesthetic and poetical portrait of the environment.

Corner's map-drawings celebrate landscape phenomena (data) in a visually poetic manner. This is evident in his map-drawing entitled "Longhouse Cave," in Mesa Colorado (14" × 16") (Figure 4.7). Corner crafts an informative image that mirrors the sun and shadow pattern of the site. The Longhouse Cave is oriented to the south but the intense summer sun enters only in the early morning. Corner draws the shape of the cave based on summer and winter sunlight throughout the day. Three images occupy the center composition of the paper: the top represents the western edge of the cave, the middle is a simple black and white plan of the cave, and the bottom image depicts the eastern edge of the cave. In the bottom right corner of the picture, Corner carefully collages a photograph of the site in a vertical rectangular shape that fits compositionally with the winter time markings. Corner's map-drawings – along with the aerial representation of the land created by MacLean – constitute a creative mode for disclosing the landscape's data. In his book, *Taking Measures Across the American Landscape*, Corner reinterprets various aerial photographs visually in the form of diagrammatic representations; he tries to reconstruct a complex system in a more logical, artistic and comprehensible way through new mapping. To quote from Corner, "The artistry lies in the use of the technique, in the way in which things are framed and set up."[29]

As in McHarg's approach, mapping is a critical part of landscape discovery. Though McHarg's approach is more scientific than poetic in its aesthetics or visual representation, today's mapping can be both an artistic and an informative process. McHarg's landscape approach is also applicable to urban conditions and is increasingly prevalent in contemporary urban landscape mapping. While definitions abound, the best way to define landscape urbanism is as a branch of "landscape ecology, concentrating on organization of human activities in the natural landscape. . . . highlighting the leftover void spaces of the city as potential commons."[30] While there are many reasons why landscape urbanism is worth mentioning, it is the potential mapping capability and visual representation of this new discipline that is most appealing. Landscape urbanism can be summed up as an arresting medley of landscape techniques. These include mapping, cataloging, triangulating, surface

4.7
"Longhouse Cave," Mesa Colorado. Drawing by James Corner.

Source: James Corner, *Taking Measures Across the American Landscape*, Yale University Press, 1996, p. 141

modeling, managing, phasing, layering and others – which can also be combined with urban design techniques such as planning, diagramming, assembling, allotting, zoning, etc. – to broaden the visual palette of the mapping field.[31]

The poetics of mapping

Corner takes a subjective approach to the objective realties of the site conditions. His map-images describe the importance of these landscapes as a productive and economical means for this territory. He also showcases the sheer beauty of this landscape, otherwise (again) left unseen. He selects pieces of the landscape through photo-collage methods and arranges them on the paper as maps. Such elements include soil conditions, climatic conditions, water and wind patterns and geometries of the landscape. His creative mapping process is a means of "finding" elements of the site, and once "found," visually presenting them in a way that tells a story of the site. As Corner once wrote, "Landscape and image are inseparable."[32] The term eidetic refers to a vivid recognizable image. Corner often uses the term eidetic in relation to landscape, as a mental conception that may be "pictureable" but as a conception that may trigger other senses: "Unlike the purely retinal impression of pictures, eidetic images contain ideas and lie at the core of processes of creativity."[33] An eidetic image is an image retained as a vivid retinal impression, an image that has left a strong impression. Corner's map-drawings stand alone as paintings; indeed, his map-drawings definitely capture the invisible through visible forms; these images are carefully composed so that they can be understood by the public and the architect with some level of instruction. It is through his poetic style and selective measures that his map-images become as they are defined – a map (the logic) and an image (the creativity). Corner's works draw the attention of the landscape architect and the general public (citizens or people interested in those specific sites profiled). He perceives the importance of having in place creative and poetic modes of representation that entertain at the same time as they inform. Maps that act as drawings or paintings present another level of the creative process, and a level that opens up what would seem to be cryptic knowledge to a broader audience. In any case, the act of contemporary mapping, as exemplified by Corner's map-drawings, is a creative process that involves critical thinking and experimental techniques of representation.

Corner sees the new trend in mapping to be one wherein poetry and sensuality are commingled. In the past, mapping dealt primarily with surveying and geography – more than ever in the field of landscape architecture and landscape urbanism, mapping is approached as an art and is often part of the design process – and should be seen as such. The mapping requires a set of choices, intelligent moves and processes.

An interview with James Corner: some final thoughts[34]

Nadia Amoroso (NA): It seems somewhat that your mapping work may be influenced by the collage techniques of the late French landscape architect, Yves Brunier, who first brought to the forefront playful and expressive collaging techniques to showcase selective pieces of the landscape?

James Corner (JC): I am not sure a reference to Brunier makes sense – maybe as a reference for comparison, as he was a collagist, but his work was in no way an influence on me personally, and may actually have appeared after the *Taking Measures* work and my earlier design work with collage and representational technique. My real references at the time (the late 1980s and early 1990s) were more Max Ernst (collage), Richard Long and Hamish Fulton (maps) and Edward Tufte (information). While the *Taking Measures* maps are well known, many of my early design drawings and collages are less published – it is probably useful to see the maps in that context, as the maps are quite technical, compared to other of my drawings and collages which are more visceral and tactile. In some ways, this tension between the technically specific and the ephemerally phenomenal has been a preoccupation of mine for many years.

NA: You take an artistic, qualitative, and perhaps even deceptive approach to mapping, in order to bring to the forefront important elements of the landscape that might otherwise be missed. You select pieces of the landscape through photo-collage methods and arrange them on the paper as maps. What is your selective process? What is the technical process (your method) in creating the map-drawings?

JC: You mention selection and choice, which is inevitable in mapping, and by extension in design. I would argue that no matter how scientific, objective and analytical the map, there is always a choice as to what is included and what is not. Maps cannot depict everything; they are inevitably partial and abstract. The assumed rationality of maps is true to the extent that they contain and convey measurable information, but at the same time there is a certain abstractness and surreality that allows maps to open up certain sets of possibilities. In this regard we can speak of a certain agency, or efficacy, that maps portend.

As landscape architects, there is an assumption that our work is rational, based upon comprehensive surveys, maps, analyses and conjectures. Design projects, like maps, are thus seen as logical outcomes of a rational process. Such rationality is desirable given the amount of financial and social investment required to transform sites. But even the most machinic, logical, linear design project contains a certain excess, a certain potential, born from selection, or from what is included and what is not. Consequently, I believe that once you understand the inevitable surrealism of strict rationality there is something very liberating in terms of how we can "play" with the world through representation. Our social constructions – maps, texts, images, projects – literally interpret and shape the world around us; they are the basis for how we see and understand the world, and as such are subject to shifts and alternative readings. Data, information, facts, are all pliable – not for the purposes of deception or whimsy (although data can easily be manipulated this way) but more for the purposes of critically disclosing new sets of possibility. Hence, my interest in the agency that certain techniques – rational techniques – have in generating new ways of seeing and creating.

You ask what my own criteria for selection are – what is my selective process. I think the answer depends on what I am looking for. But generally, the process tends to be both rational and accidental. It is rational to the degree that I assemble and review as much information as possible, extracting insights that I believe are most central to the issue at hand. But it is also accidental to the degree that what I discover in the process might actually condition what I end up believing the issue at hand actually is. This may go so far as to completely reverse and contradict what I thought the issue was originally. Thus, the process of assembling information, reviewing different data sets side by side, sometimes in random and disconnected adjacencies – as in two images out of a huge pile accidentally sitting right next to one another, creating the metaphoric link – can quite often lead to the construction of a very particular way of seeing and understanding, both rational and personal. Techniques such as piling, collating, sorting, mixing, overlaying, flicking, editing, and so forth, point to a process of digging through material to uncover hidden potentials, ideas, relationships and possibilities. It is rational, accidental, critical and speculative, all at the same time.

NA: What is the relationship between scale and your graphic type? Then, there is also the issue of scale accuracy – is this relevant?

JC: Scale is interesting, and again parallels the rational/speculative action I am describing above. Scale is rational and measurable, and allows for the most effective translation between representation and action – such as when we use drawings to construct a building, or when we use a map to find our way to a particular destination. But scale is also speculative, allowing for potential misreadings that alternate between the miniature and the immense. One can easily get lost in a map.

With regard to misreading scale, it is possible to take a simple black and white composition on paper, and to imagine that composition to be a "to-scale plan" of a building or landscape, for example. Depending on what scale you set, the plan could be referring to a furniture layout in a room, to a large garden, to a regional landscape. The imagination can play very effectively with scalar shifts and varied interpretations of what one is actually looking at. Thus, scale is best understood in this double sense, as something both rational and scaleable on the one hand, and both re-scaleable and un-scaleable on the other. Different people have played with this principle in a variety of interesting ways – Peter Eisenman, with Laurie Olin, for example, developed a practice of geographical map scaling in the early 1990s, that led to some fascinating projects (Cannaregio, Venice; or Long Beach Museum, for instance); Richard Long alternates scale in his map/walk itineraries, and invokes both durational scales of time alongside length and space; Philip and Phyllis Morrison's "Powers of Ten" suggests a sort of nested interrelatedness across scales; Deleuze and Guattari speak of maps and scales in very suggestive and imaginative ways; and I would argue that a landscape architect such as Le Notre was deploying multiple scales of geometry and content in his work. So, scale is an extraordinarily important and fascinating subject – impossible to escape in landscape architecture, and potentially one of landscape architecture's most potent creative tools.

NA: Do you think your map-drawings are tools for designs or tools for making better design decisions?

JC: The maps in *Taking Measures* were more intended as speculative depictions of geometrical and measured content related specifically to certain landscapes across the United States. At the time, I was interested in how we could better understand form, geometry and scale as a result of very specific and often quite pragmatic land settlement practices, and the role that surveys, maps and images played in the marking out of the American landscape. Consequently, the map drawings in the book are not really designs or tools for design, and they are not necessarily useful for making decisions. They are simply intended to reflect on how maps and measured aerial depictions (which are typically flat, planar and horizontal) describe certain conditions alongside more scenic representations (Alex McLean's oblique, perspective photographs).

On the other hand, it is possible to extrapolate from this work how maps, surveys, data and collagic techniques of adjacency can in fact serve as design techniques. The inevitable correspondence between representational technique and design production has been well documented and remains a major area of research and interest in the field. Plans and perspective views, for example, offer two completely different ways of seeing, thinking through and organizing space. Maps, too, offer significant differences in both content and viewpoint, depending on the map projection system, scale, codes and information. Their geometrical construction can also be of enormous significance for design interpretation and projection. So, yes, to answer your question, maps are fundamental to design creativity, instrumentality and content. Their efficacy in any design process is dependent upon how creatively the mapping game is extrapolated.

NA: Ian McHarg has introduced scientific and ecological mapping systems that were critical to landscape architects then and to some extent now. In some respects, you have brought landscape mapping to the contemporary visual and cultural forefront for landscape architects to understand specific measures of the landscape through the map-drawings. Can you please comment on your thoughts on future trends relating to spatial mapping in landscape architecture with respect to visual communications?

JC: There will always be maps in landscape architecture. It is inconceivable that a landscape architect could meaningfully approach a project site without the information provided by maps and surveys. McHarg's belief was that a simple topographic site survey was inadequate to fully understand the ecological and historical forces across any site. He required the collation of separate maps showing in detail a specific site's climate, geology, geomorphology, contour, soils, hydrology, vegetation, habitat, flora and fauna, historical patterns of land use, socio-cultural histories and economic factors. This information-dense "layer-cake" of different components allowed for a more ecological and dynamic understanding of how a site came to be and how it may be most sustainably modified. For McHarg, site design was not simply a matter of spatial shape, form and experiential impression but more the adaptation of various ecological forces in time to shape a more dynamic and complete whole. And of course maps and mapping techniques were fundamental to the McHarg project. I believe that this is just as valid and important today.

The only difference I would add to the McHargian view of mapping is the promotion of individual creativity. McHarg sought both comprehensive inclusion and measurable objectivity in his mappings. I endorse this, but at the same time I allow for – and embrace – the constructed and abstract character of the mapping project. When Richard Long takes a map and inscribes an itinerary upon it to then follow, he is not simply using the map as a neutral data base but is extrapolating a new project upon it, a project both willful and determined. McHarg's maps were supposed to produce outcomes, conclusions, whereas I believe, like Long, that maps are instruments to be willfully and critically engaged. In this sense, maps are both a neutral data source (you look at them) and a basis for significant action, for physical manipulation of the data (you work with them).

In this regard, you ask about current and future trends for mapping. There is of course much to be acclaimed in modern Geographic Information Systems, which use computers and complex programs to assemble and sort through spatial and quantitative data sets, often to study alternative development scenarios and to assess strengths and weaknesses of one approach over another. This is good and important work. But I believe another body of research should look at how maps are inevitably cultural constructs, not simply inert rational data banks, but active diagrams that extend a certain agency over how the world gets shaped. Artists and conceptualists are good at seeing maps in this way, not so much as informational devices, but as performance stages that can critically script certain spatial geographies in fresh ways. I believe that if we can develop a better understanding of maps as both data resource and alternative viewing and scripting device, then we will have found a significant new tool in the arsenal of creative agency.

Notes

1 Edward Tufte, *The Visual Display of Quantitative Information*. Second Edition, p. 79.
2 Ibid., p. 177.
3 Ibid., p. 77. A Summary of Tufte's principles for graphical integrity can be located in this reference page. These are also profiled in Chapter 3.
4 Foreword by Michael van Valkenburg, *Taking Measures Across the American Landscape*, by James Corner and Alex MacLean, p. ix. Van Valkenburg is a graduate from Cornell University and a Professor in Practice in Landscape Architecture at Harvard's Graduate School of Design. He has helped refine and advance landscape architectural education and has promoted creative modes of visual communication in the field.
5 Corner, *Taking Measures Across the American Landscape*, p. xi.

6 Ibid., p. xix.
7 Ibid., p. 18. "Aerial Representation and the Making of Landscape."
8 Merriam-Webster Online Dictionary, "Agency," http://www.m-w.com/dictionary.htm (accessed December 2007).
9 James Corner, "The Agency of Mapping: Speculation, Critique and Invention" in *Mappings*, edited by Denis Cosgrove, p. 216.
10 Ibid., p. 213.
11 Center for History and New Media, George Mason University, "Maps: What Makes a Map a Map?" 2003–2005. http://chnm.gmu.edu/worldhistorysources/unpacking/mapswhatmakes.html (accessed May 2008).
12 Ibid.
13 Center for History and New Media, George Mason University, "Maps: Why Bother with Maps?" 2003–2005. http://chnm.gmu.edu/worldhistorysources/unpacking/mapswhybother.html (accessed May 2008).
14 Transcript of Professor Gerald Danzer's lecture on maps, obtained from Center for History and New Media, George Mason University, "Introduction: Analyzing Maps." 2003–2005. http://chnm.gmu.edu/worldhistorysources/analyzing/maps/analyzingmapsintronf.html (accessed May 2008).
15 James Corner, "The Agency of Mapping: Speculation, Critique and Invention" in *Mappings*, edited by Denis Cosgrove, p. 215.
16 Julia Czerniak, "Challenging the Pictorial: Recent Landscape Practice" in *Assemblage*, 34, 1998, p. 110.
17 Ibid., p. 111.
18 Ibid.
19 Ibid.
20 John A. Jakle, "Reviewed work(s): *Taking Measures across the American Landscape* by James Corner, Alex S. MacLean" in *Annals of the Association of American Geographers*, 87, 3, Sept. 1997, p. 538.
21 Corner, *Taking Measures Across the American Landscape*, p. xvii.
22 Ibid., p. 125.
23 Ibid., p. 19.
24 John Jakle, "Reviewed work(s): *Taking Measures across the American Landscape* by James Corner, Alex S. Maclean" in *Annals of the Association of American Geographers*, 87, 3, Sept. 1997, pp. 537–538.
25 James Corner, "The Agency of Mapping: Speculation, Critique and Invention," p. 213, in *Mappings*, edited by Cosgrove. This publication profiles historical and contemporary accounts pertaining to issues of mapping, maps, symbolic landscapes, and visual communication in cartography. Academics and practitioners in these fields or related fields who have contributed to this excellent publication include Denis Cosgrove, Christian Jacob, Alessandro Scafi, Jerry Brotton, Lucia Nuti, Michael Charlesworth, Paul Carter, Luciana de Lima Martins, Armand Mattelart, David Matless, James Corner, and Wystan Curnow.
26 Ibid.
27 Ibid.
28 Ibid., p. 215.
29 Ibid., p. 251.
30 Grahame Shane. "The Emergence of "Landscape Urbanism": Reflections of *Stalking Detroit*," in *Harvard Design Magazine: On Landscape*, Fall/ Winter 2003, p. 4. The term "Landscape Urbanism" was coined by Charles Waldheim, presently Director of Landscape Architecture at the University of Toronto, in a discussion with James Corner in 1996. Landscape urbanism focuses on the notion of "landscape as urbanism." It examines the emerging role of landscape as a means and medium for architecture and urban design,

exercising landscape as organizing both horizontal and vertical conditions, in essence, surface conditions.
31　James Corner, "Landscape Urbanism" in *Landscape Urbanism: A Manual for the Machinic Landscape*, edited by Moshen Mostafavi and Ciro Najle. London: The Architectural Association, 2003, p. 62.
32　Corner, "Operational Eidetics: Forging New Landscapes" in *Harvard Design Magazine*, Fall 1998, p. 22.
33　Ibid.
34　An interview with James Corner of Field Operations was conducted in August 2009.

Part II

Visuals

5 Drawings

The Map-Landscapes

The creative map

Crafting maps as both aesthetic objects and as empirical evidence is important – *their form, the composition, artistic license, the capturing of the information, the spatial relationships between form and data, the color application and lighting, and their metaphorical relationship between the data type and presentation.* The quest to create spatial maps in relation to the unknown city – *maps that can serve as design tools and as examples of art* – is profiled in this chapter. Today, thanks to the efforts of some forward-thinking individuals, architects and urban designers are more receptive than ever before to using alternative modes of representation to unravel the cityscape's *deeper* issues.

How can the abstract forces (urban phenomena) shaping urban life be rendered *artistically, spatially and informatively* in the form of alternative "maps" which represent urban dynamics – urban dynamics not usually accessible to urban designers or the urban dweller? The balance between the aesthetics (the art) and the empirical evidence (the data) is fused together to establish a new breed of multi-dimensional maps, termed "map-landscapes." These hybrid maps combine the creativity, the logic and the spatial qualities, revealing the invisibles of the city. They are rendered in a manner that offers deeper insights into the urban space. These map-landscapes become tools which can steer and influence the urban design process. They grant additional readings to the city space inspired by data and driven by a creative force, similar to the *Evolution* drawings by Ferriss. Using new and alternative modes of graphic display, these maps express otherwise unseen urban data – "exposing" the city. An image of the city that has artistic merits attracts its viewers with its seductive qualities and is likely to inspire immediate reactions. In an attempt to inspire better built forms, new modes of representation are proposed. Gauging the extent to which a scheme is capable of manifesting those invisible urban forces which define the essence of a city also determines the usability and value of its mode of representation.

In short, aesthetically pleasing mapping engages the cognitive faculties of the onlooker to an extent that perfunctory and visually unappealing mapping does not. Imagine manifesting the city's phenomena – "invisible forces" including land values, areas under public surveillance, hours of operations, population density, etc. What shape will they take? With the new urban forms represented as images, would they become guides into the city in the form of new maps? Could these images provide a means for more intelligent urban design by revealing the city's hidden potentials? How will they be manifested?

This chapter illustrates a collection of map-landscapes for selected global cities – Toronto, New York and London – to test the urban spatial parameters perceived by the urban "invisibles." These visual experiments are shaped by certain guiding principles as well as by historical and contemporary works. Also profiled are creativity and sculptural maps produced by a group of graduate students at the University of Toronto and at Cornell University. Raw data for the maps of Toronto, New York and London need to be drawn from censuses, city planning offices, police departments and other published sources. The maps are composed within a field-set (set of parameters) in which the quantifiable data are examined and analyzed for visual abstraction. These images are new maps for the twenty-first-century metropolis – ones that guide us into the unknown through a new form of representation – and inspired by the works of Ferriss, Corner, MVRDV, Tufte and Wurman. How can textual information be rendered into a "poetic graphic," which offers value for both the architect and urban dweller? Free from legends, text, hard line drafting conventions, and traditional mapping methods of the past, these maps provide an image of the city as it truly is, based upon urban statistics. The images or map-landscapes profiled in this chapter reveal snapshots of the living city. The visual connective tissue between the potent unseen forces that shape the urban environment and the physical embodiment of the urban environment are expressed in them. In essence, they expose the "unacknowledged" elements that affect this landscape and its surrounding urban environment. We, as a society and in the field of the arts, are interested in beauty, the visual appeal. The process transforms policy, regulations, and urban activity and other factors into the multidimensional.

Throughout the visual experiments, one additional aim is to show how digital media can lay bare the true nature of urban form; technology, in this instance, can be our friend. In any event, the dynamic phenomena visualized include socio-political aspects, such as the dynamics of real-estate development, crime rates and building envelopes. Other dynamic phenomena take shape through indicators such as air-quality readings, density patterns, global patterns, business hours of operations and number of people occupying certain entertainment establishments and traffic volumes. Some zoom in to a specific district of the city to capture a finer profile.

The balance between the artistic and informative qualities in the creation of the map-landscapes can be summarized as the following: Do the drawings convey clearly the urban data they try to represent? Does the audience understand the type of urban information through the representation? Are these images new guides to the city? Do the images have artistic merits and aesthetic qualities (balance of light and shadow, composition of subject matter, color, technique, sculptural)? Do the images attract the audience in a way that conventional maps do not? Do they provoke an immediate reaction? Would you display these images as works of art? Do the maps evoke a sense of form or spatial qualities? Is there an act of discovery? What medium is selected to powerfully expose the city's unknowns?

Referring back to Ferriss's maps/drawings, his works seduced a range of audiences by the mere style and selected data represented. City officials, architects, citizens of New York – all found his work compelling. Using charcoal as his medium of choice, Hugh Ferriss rendered artistic interpretations of the Ordinance through his chiaroscuro techniques. The drawing medium used both for creating a powerful visual effect and communicating an architectural vision establishes a particularly strong link between the message of the drawing (the Ordinance) and its technique of production (the art). Scanning the Ordinance, it is clear that Ferriss's maps are encoded with many layers of technical and abstract data reflecting the legal, economic, and political circumstances within Manhattan; it can be said that Ferriss's maps produce both a precise and an imaginary fusion of the present and, at times, future conditions within New York City. Ferriss's maps not only express a possible objective reality, but are also charged with his emotional input.

In Wurman's current work on methods of drawing and mapping that can make urban information comprehensible, he states that it is absolutely necessary for information about the super-cities of the world to be presented understandably and comparatively. In Tufte's work, his lifetime goal is to review and to create principles for visualizing information. Tufte and Wurman are considered experts in the field of information graphics; they have both sought ways to make the complex clear and to represent information in an understandable manner. In 1966, Joseph Passonneau, an architect, civil engineer and academic, working in conjunction with Richard Wurman, created the publication, *Urban Atlas: 20 American Cities: A Communication Study Notating Selected Urban Data at a Scale of 1:48,000*. In this publication, Passonneau and Wurman profiled twenty American cities, including Boston, New York and Washington, DC, and created maps depicting urban information for them. Employing old technologies, Passonneau and Wurman visually documented population density, using dots and circles to indicate the number of people that occupied given areas (Figure 5.1). Plotting and mapping the information by hand was a laborious and inaccurate task. Working by hand, these two architects created an urban atlas that depicted population density, income intensity (personal income) and land use. Passonneau and Wurman used circles, dots and squares (unfilled and filled) within a fixed grid set, with each depicting the value of information he was trying to visualize. It may sound like a primitive mode of representation compared with

RESIDENTIAL POPULATION DENSITY   ○ 50·200   ◉ 201·500   ● 501·1200   ● 1201·3600   ● over 3600

5.1
"Residential Population Density, New York," a two-dimensional map created by Passonneau and Wurman. Similar mapping renditions are found in Passonneau and Wurman's *Urban Atlas* (1966) depicting various urban data sets for twenty American cities.

Source: Joseph R. Passonneau and Richard Saul Wurman, *Urban Atlas: 20 American Cities: a Communication Study Noting Selected Urban Data at a Scale of 1: 48,000*, 1966

modern digital methods, but in the 1960s when this publication was produced, their technique was highly inventive.

Passonneau and Wurman's book became a seminal text in the field of urban visual depiction. In it, they put forward the following propositions:

1. It is the work of the urban architect to capture, in a geometric web, such varied and dynamic human and natural elements. 2. The form of a city and the forms of its many elements are therefore shaped by multiple, interdependent forces, and each has independent measures of excellence. 3. Urban form carries a surcharge of information that can be intuited or "read" by people whose history and emotional and intellectual background make such information accessible to them. This is why we often speak of the "language of architecture" or the "language of form."[1]

Densityscape maps

Figures 5.2–5.28 are by the author. The *Densityscape* maps expose new urban forms based on population density (people/unit area). Created using software programs including 3-D MAX, 3-D Viz and Photoshop software, information is transformed into

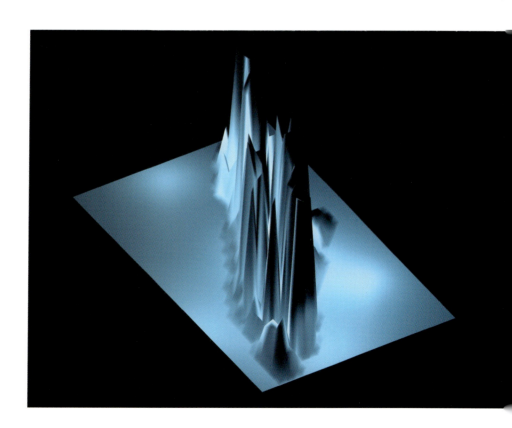

5.2
"Densityscape, New York City," a three-dimensional map depicting the super-dense residential population landscape of Manhattan Island.

Drawings

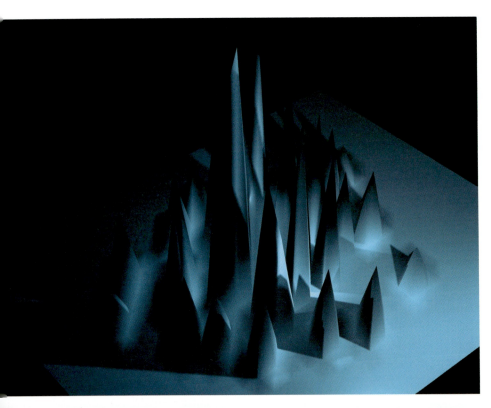

5.3
"Densityscape, London,"
a three-dimensional map
depicting the dense, yet more
dispersed residential
population landscape of
London.

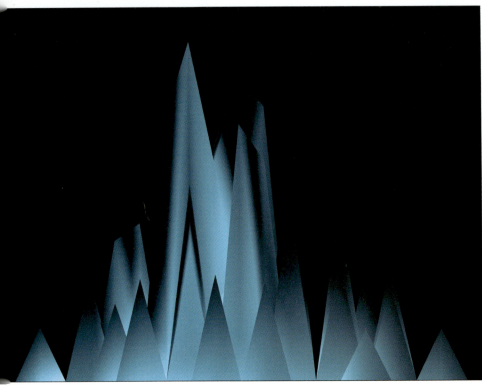

5.4
"Densityscape, London,"
sectional view of the
map-landscape, looking
north.

a landscape – through a digital process – taking a flat surface of embedded information, and tapering, folding, pushing and/or pulling this surface based on value systems.[2] The plane is occupied by a tight grid of flexible bands,[3] which can be manipulated into a new form depending on the input values. Slowly a new city envelope is created. An honest and compelling interpretation of urban statistics is brought forth through these new configurations. One of the first new maps profiles New York City. What would New York City look like spatially if it were perceived by population per district? What kind of new city form or landscape would be generated? Figure 5.2 is a portrait of New York City characterized by population density. As an image, Manhattan (the subject) is strategically positioned in an isometric view looking north-east. The correct balance of light was used to highlight specific peaks and crevices of the form, crafted using statistics as the design palette. The map "Densityscape, New York City" has some resemblance to the dense New York urban mass.

The positive "pullings" were used in the creation of the population density maps; the topographical interpolations of this information are literally translated into "peaks" and "valleys" that are instantly comprehensible to most viewers. The flat ground (surface plane) is assigned a value of zero for all the maps to indicate no-value or an untouched or unaffected area.

Architect and professor Rodolphe el-Khoury comments on my images:

> It's helpful in the sense that architects or urban designers usually can react immediately to a context by looking at the topography. You know they see hills, they see ravines, they understand that immediately and they can react to it. They see buildings they can comprehend in a very direct way, but there are invisible forces in the city that are not so intelligible right away and that merit equal attention, so these kind of drawings actually will make them much more tangible, more accessible so that designers can react to them just the way they would react to a hill or to a building, so those forces become much more influential in their designs.[4]

In the map created for London's population (Figure 5.3), one sees a diverse topographical landscape; the light, shadow and composition all become elements which create a portrait of the city's population density; moreover, the deep shadows complement the sheer height of the tallest peaks which mark the Boroughs of Kensington and Chelsea, Islington, Hackney, Hammersmith and Fulham, Lambeth, Tower Hamlets and Camden. When viewed in a profile section (looking north), the image has an alternative reading (Figure 5.4) and allows one to see how the centre of town is the area where the greatest population density is to be found.[5] Visually, the light, shadow and the rhythm of the peaks are carefully composed and balanced as an artistic piece. The map "Densityscape, London" has far fewer points of interpolations (thirty-three in total). Each point is located in the most "central" spot of the

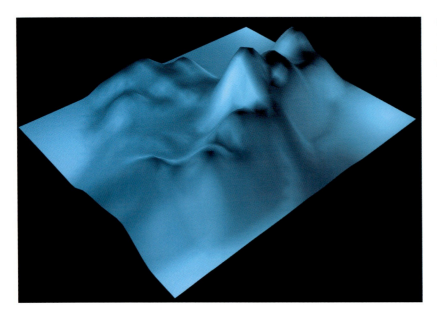

5.5
"Densityscape, Toronto," a three-dimensional map depicting the smoother, more hill-like residential population landscape of Toronto.

5.6
"Densityscape, Toronto," a sculptural map-landscape crafted from the digital form using CNC (Computer Numerical Control) process.

5.7
"Densityscape, Toronto," a sculptural map-landscape with aerial photo of the City of Toronto projected onto the form for referential purposes.

defined boundary for each Borough. Based on the selected interpolation points, one can see that London is still a fairly dense city but has more land area to "contain" the number of people per district as compared to the "super" dense map "Densityscape, New York City."

Taking the data for population density, the same approach is taken for the City of Toronto (Figure 5.5). Here a much more "hill-like" picture is created. The vertical scales used are the same for all three cities. When the images of Toronto, London and New York are displayed at a gallery exhibition, one can immediately see the difference in population density between these three global cities. Informationally speaking, these maps capture the dense character of the city. Each map is carefully positioned in an isometric view to receive an optimum viewpoint of the peaks and crevices of the city portrayed. One sees this image looking north-east and the onlooker is struck by the highlights on the valleys and hilltops in the composition. Using a CNC (computer numerical control) fabrication process, it is possible to craft a sculptural piece from the compositional information in the Toronto density maps (Figure 5.6). This piece evokes a much more emotional and intuitive response. A drill bit carves the piece of wood. Each precise cut allows the mapmaker to carefully examine the high and low points on the wood which directly link to values for population density. Now the map is a sculptural piece, and an artifact.[6] Much like Walter Kilham's plasticine model demonstrated the maximum envelope of the Daily News building (1927), this sculptural map begins to spatially render the outcome of the city's landscape based on this particular data type. When hung on a wall with an aerial photograph of the city projected onto it, the viewer begins to draw connections between the bumps (height points in the sculpture) and the aerial photo (Figure 5.7).

Crimescape maps

These maps define new urban forms for the cities of New York, London and Toronto through the incidence of crime. By contrast with the density maps, these sets of maps are generated by negative "pullings": showing areas of high crime with high values in the negative z direction. The reason for this is that crime is associated with negative outcomes. Traditionally, crime maps use district or area color codes to indicate crime numbers. With the use of the internet, one can obtain textual data on crime by selecting, for example, the Boroughs of London.[7] Boroughs with the highest crime levels include Westminster, Lambeth, Southwark and Newham.[8] In any case, the image received is two-dimensional and not volumetric. In the image "Crimescape, London" (Figure 5.8), the clever use of dark grayish colors makes a direct allusion to the notion of crime, while light intensity is used to highlight areas of greater crime; the light and shadow dramatize the haunting character of the subject matter that this map is trying to represent, especially when a snapshot of the London Crime map is taken in sectional or profile view (Figure 5.9). Allegorically, the image resembles icicles, hinting at the notion of coldness, darkness and eeriness. The negative "topographical forms" of the city are depressed with deep pit-like

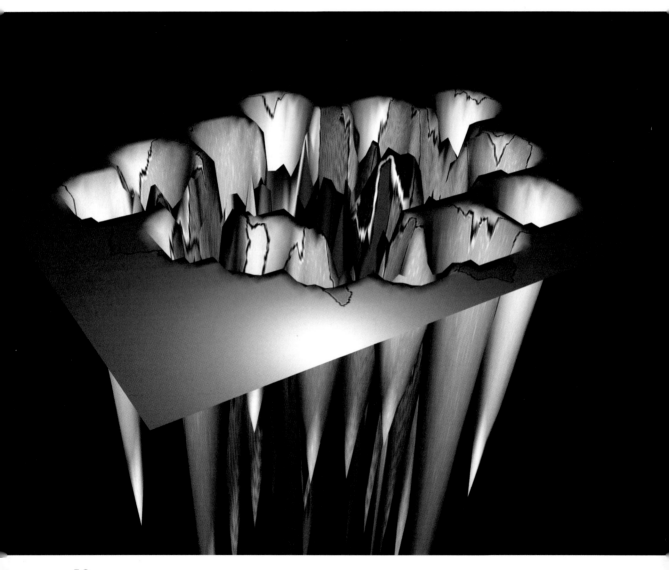

5.8
"Crimescape, London," a three-dimensional map with negative "pulling" showing a "depressed" sharp pit-like landscape formed by crime indices.

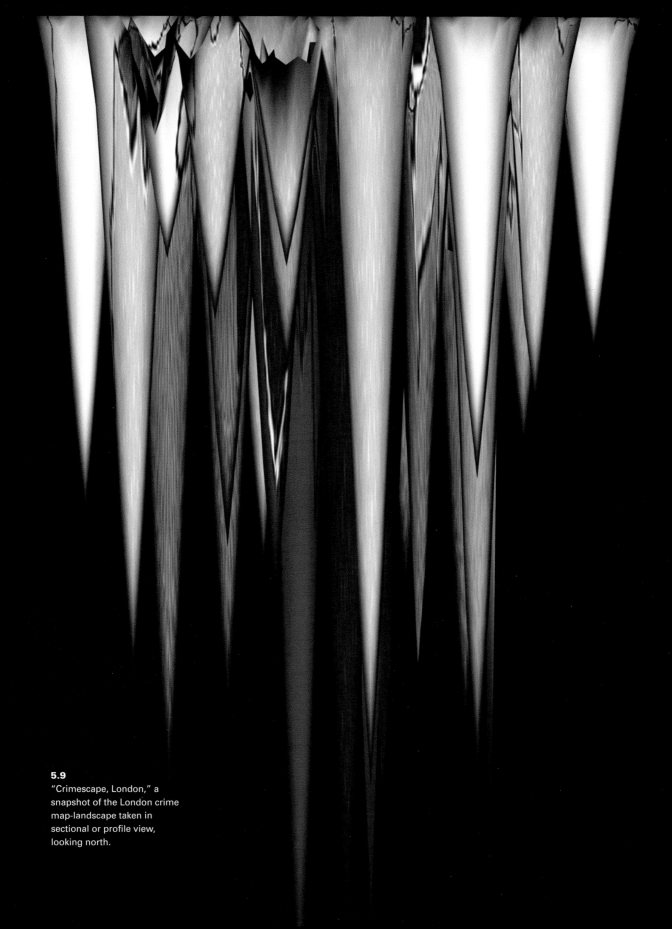

5.9
"Crimescape, London," a snapshot of the London crime map-landscape taken in sectional or profile view, looking north.

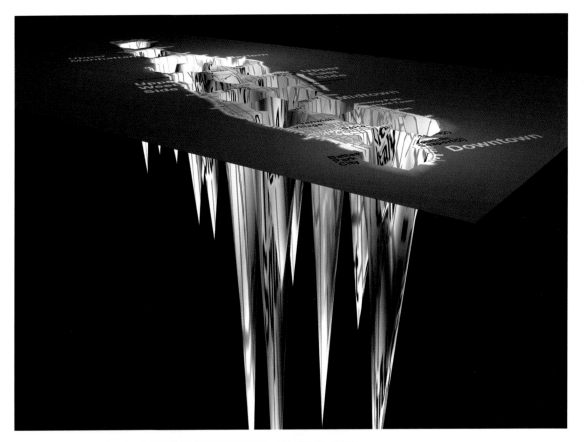

5.10
"Crimescape, New York City,"
a map-landscape showing a
crime-identifying depiction of
New York City.

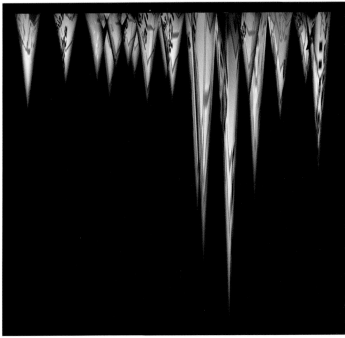

5.11
"Crimescape, New York City," a snapshot of
the New York City crime map-landscape
captured in sectional view looking east.

points to create a crime-identifying portrait of New York (Figure 5.10). These sharp, pit-like landscapes relate to a hellish underground landscape which presents a clear and provocative underpinning of this particular characteristic of the city. Once again, the angle of view of the map and shadow and lighting are all contributing factors in creating a work that is artistic as well as enlightening. In Figure 5.11 (a section view of New York), the icicle-like image presents an uncanny representation of the subject matter. Artistically, these images are visually striking, capturing the audience's attention. When on display in a gallery, the urban dweller comments on these images as digital paintings, using new media to capture visually the potency of the elements that affect our city.

In the Toronto *Crimescape* maps, the new urban form of the city is better understood when the phenomenon of crime is taken into account. The image is again positioned in an isometric view. A line drawing of the Toronto districts is placed over the crime rate map-landscape (Figure 5.12) and this gives the audience an immediate understanding of the relation between criminal activity and a specific geographic location. The sectional view of this "Crimescape, Toronto" conveys images of "fangs" or "ferocious teeth" (Figure 5.13), alluding to the negative conations of crime. This urban portrayal depicts Toronto's crime as "a hellish underground of fierce spikes of criminality pointing straight down,"[9] they show us the areas within Canada's largest metropolitan community that need special attention from law makers.

Market-valuescape maps

These images express the new urban form created by land value information. The districts or boroughs are the land value areas. The data are mostly associated with large areas – which produces spiky results. Small areas of data would give a smoother, more hill-like surface. Figure 5.14 expresses a new landscape derived from market value data for Toronto, based on the interpolation of the surface across the data points, which creates a smoother surface. Green is the selected color to depict this landscape. In this case, a mesh-like rendering is chosen to express the structure of this landscape. In "Market-valuescape, New York City" (Figure 5.15), a green texture is applied and a street map added to help guide the viewer in terms of location. In this case, the landscape form has some labeling and text identifying specific locations. With the placement of text, the map gives the viewer a more direct relation to space, whereas the Toronto market value form is more abstract. The New York form becomes less abstracted and less artistic, and becomes more self-evidently informative.

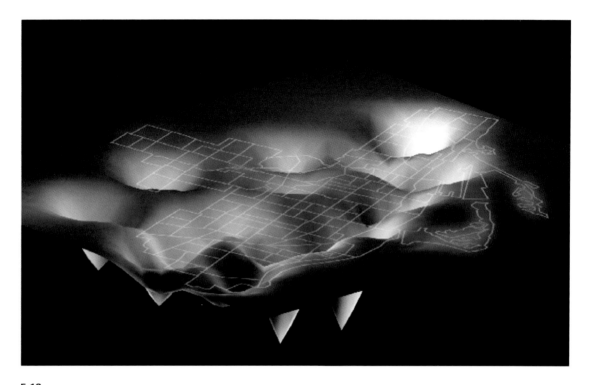

5.12
"Crimescape, Toronto," a three-dimensional map with negative "pulling" showing a negative topographical landscape formed by crime indices. A simple line drawing of the Toronto districts is superimposed over the crime rate map-landscape.

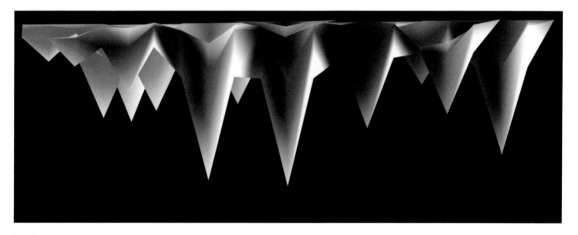

5.13
"Crimescape," sectional view of the Toronto crimescape, looking north, conveying images of "fangs" or "ferocious teeth," suggesting the negative connotations of crime.

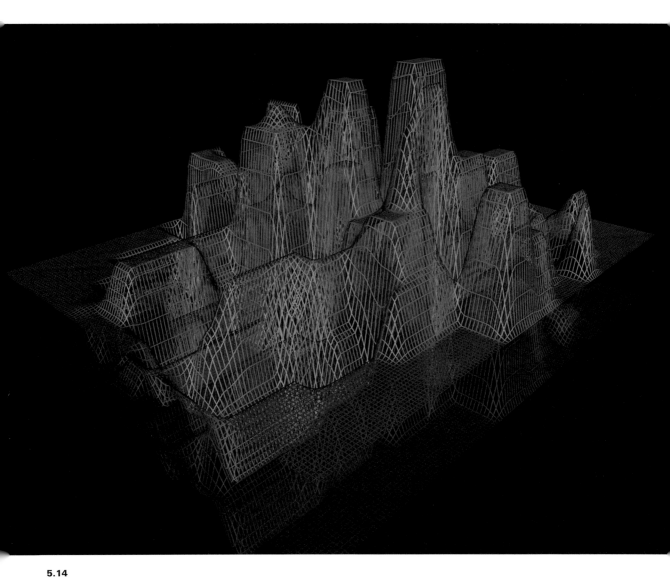

5.14
"Market-valuescape, Toronto," expressing a new landscape derived from market value data, based on the interpolation of the surface across the data points, which creates a smoother surface.

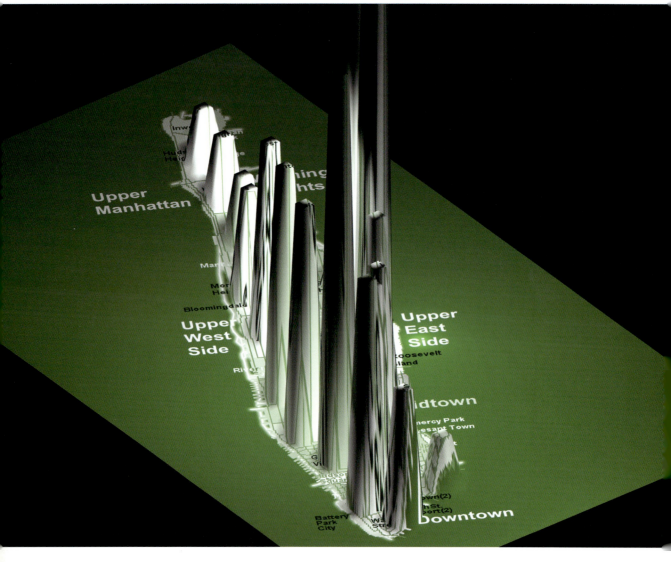

5.15
"Market-valuescape, New York City," a much more mountainous-like landscape generated based on higher market values in the city. This map-landscape is comparable in form to its "Densityscape" map-landscape.

Valuescape

Figure 5.16 is a sculptural piece that showcased the market value findings for Toronto. Different from the CNC piece depicting density levels in Toronto, this piece was based on an old punch card system. Plastic rectangular plates inscribed with a line map of Toronto contained holes of three sizes (small, medium, and large). A series of identically engineered plastic pins were created to fit into plates with similar hole sizes. The smallest hole size would accommodate the smallest diameter pin, resulting in a shallow depth settlement in the card and a higher peak; the largest

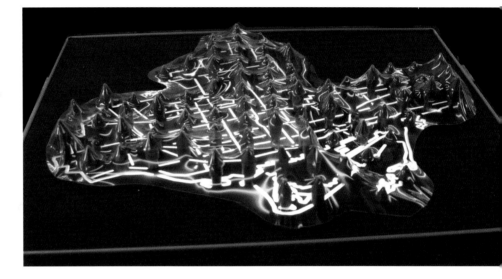

5.16
"Valuescape, Toronto," sculptural piece that showcases various value points of the city registering "small, medium, and large" market values. A clear plastic sheet is draped over these points and then vacuum sealed to produce the oddly pin-like landscape.

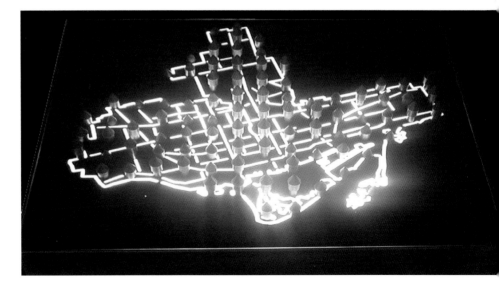

5.17
"Valuescape, Toronto," sculptural piece that showcases a forest of dark grey pins depicting a landscape of market value which is illuminated at its base.

Drawings

hole sizes relate to the lowest assessment in market values while the smallest hole sizes relate to higher assessments of land market value. Once they are all dropped and settled into place, a forest of dark grey pins depicts a landscape of market value which is illuminated at its base (Figure 5.17). When there is a change in market values, a new card reflecting these changes is created. The pins settle at a new level, creating a new landscape of pins. To capture the ever-changing forms, a clear plastic sheet is draped over these pins and then vacuum sealed to produce the oddly pin-like landscape; this clear sheet then forms its own surface. These clear plastic sheets can subsequently be removed and displayed as landscape objects. The crystallized

sheets can also overlay each other and allow individuals to see a difference between one market value landscape and its counterpart. Also, the inscribed Toronto district map on the plate provides further "cognitive connections" to locations within the city.

Surveillance-scape maps for Toronto

These map-landscapes are perceived by public surveillance cameras situated in the Financial District of downtown Toronto. "Conal Abstraction, Toronto" (Figure 5.18)

5.18
"Conal Abstraction, Toronto," a new landscape surface generated by public surveillance cameras in a downtown core of the city.

portrays the volumes within the fields of vision of surveillance cameras. Intersecting cones of vision become intensely lighted, allowing one to understand that this is a physical or geographic space in which one is mostly likely to be viewed. The bright red is a powerful hue which is meant to attract the viewer. According to Peter Goddard:

> "Conal Abstraction, Toronto" – in which megaphone-like shapes intersect gently to represent downtown surveillance cameras – could well have been part of the work shown at the Salon des Independents cubism show in Paris, 1911.[10]

Figure 5.19 is the same rendition of public surveillance, but the gray transparent forms suggest non-physical force, suggesting a phantom city based on public surveillance. The points of the cones are generated from the point of surveillance. Intensely lighted areas depict overlapping camera views, where in the city one is most likely to be watched by an unblinking eye.

5.19
"Surveillance-scape, Toronto," a snapshot of public surveillance in which the gray transparent forms suggest non-physical force.

Traffic Flux maps

Figure 5.20 shows a sequence of still frame snapshots of Manhattan's traffic movement over a 24-hour cycle. Based on the analogy of a heart, rhythmic images of a pulsing network of red tubes can depict the traffic flux of this dynamic city. A collection of red tubes woven together to trace the road network of New York City portrays them expanding and contracting, based on the number of vehicles that occupy each road section. (This was part of an animated map series which constantly shows the rhythmic flows of activity on a 24-hour cycle. The growth or ebb and flow of these expanding tubes help define the image of the city.) Please refer to www.nadiaamoroso.com to view the animated maps.

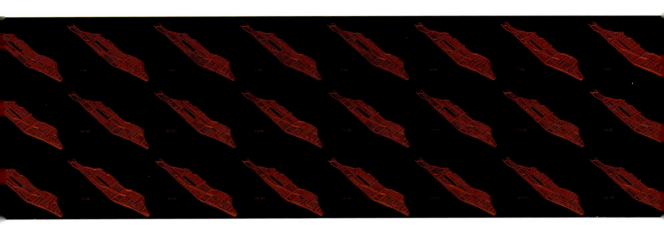

5.20
"Traffic Flux, New York City," showing a sequence of still frame snapshots of Manhattan's traffic movement over a 24-hour cycle.

Ozone Space maps

Turning to London, a new urban form or "living map" which shows the city's virtual space of ozone levels in different Boroughs[11] is created. A series of twenty-four snapshots depict bright red expanding and contracting spheres that correspond to specific ozone spaces. This was part of an animated map showing ozone level changes in a 24-hour cycle. In essence, ozone is treated as space. The center point of the bubble "grows" from the central point of the Borough. The volume of the sphere is proportional to the ozone volume. However, if the ozone reading is high, the bubble expands beyond the Borough boundary lines and often intersects with other ozone "spaces." As a spatial map, this may suggest specific areas of environmental concern.

The sphere size shows the quality of ozone in the city; if the level of the ozone is dangerous, the size of the sphere grows; if it is low, the sphere diminishes in size. Figures 5.21 and 5.22 depict snapshot images of various ozone levels (μg/m – micrograms per cubic meter of air) recorded in London at 8.00 a.m. and the other at midnight on a given day in June 2001 (seen in slight aerial and sectional views). Please refer to www.nadiaamoroso.com to view the animated maps.

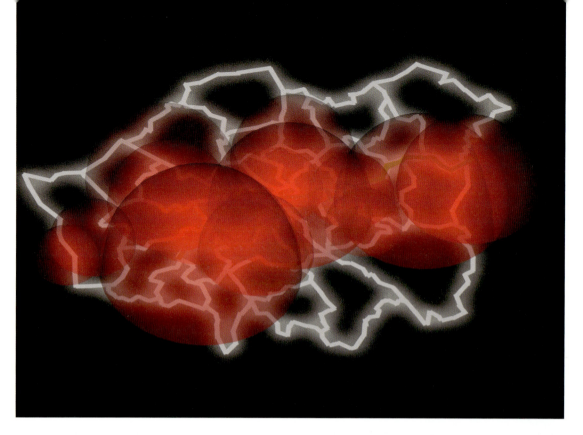

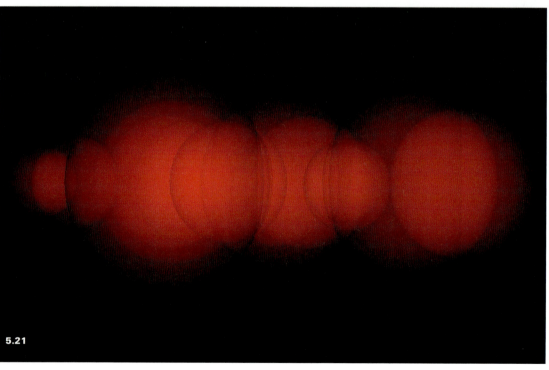

5.21

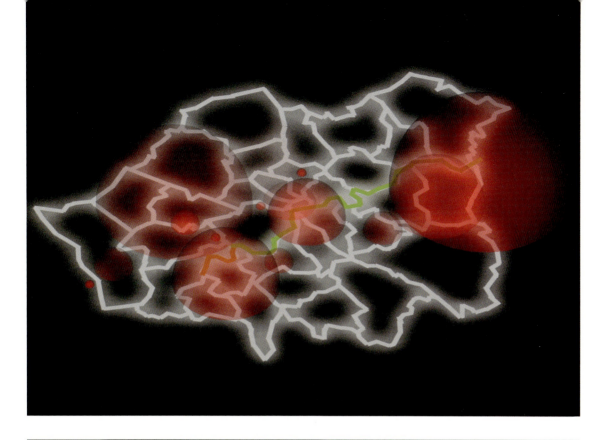

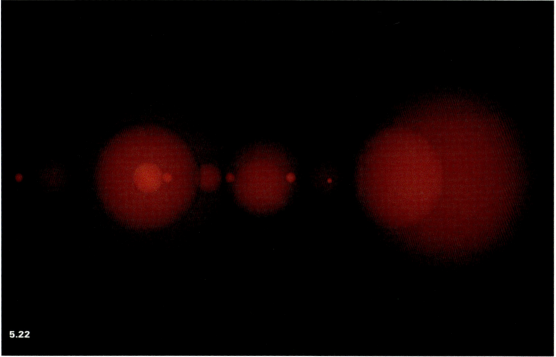

5.21–5.22
"Ozone Space, London," three-dimensional map-landscapes depicting ozone levels at various times of the day, with accompanying sectional views looking north.

Air-quality Index maps

Figure 5.23 depicts air-quality levels on a given August day in Toronto. The darker red, more opaque bands communicate poor air-quality levels, while those which are transparent and brighter indicate better air quality. Each band is layered on top of the previous one, indicating various times of the day and air-quality reading. Morning hours are registered at the bottom outline and the top outline indicates evening hours. The outline of Toronto serves as a base map and indicates the air-quality location for Toronto east, west, north and south. Though the images are created digitally, they have a painterly quality; for instance, the chalk sketched white line trace of the Toronto map is smudged along the ends to strengthen the artistic quality of the map.

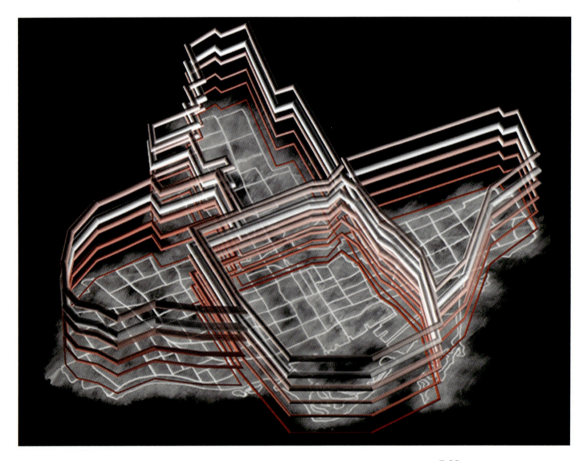

5.23
"Air-quality Index Map, Toronto," using bands of various thickness and color intensities to register poor air quality at various times of the day.

Zoning By-laws maps

These map-landscapes capture zoning "forms" for a downtown block in Toronto. Inspired by the works of Hugh Ferriss, the images are crafted in an evolutionary process much like Ferriss's staged drawings. The first stage captures raw bulk height

5.26

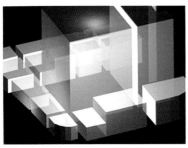

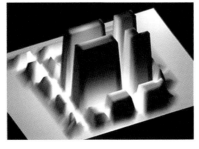

5.24　　　　　　　　5.25

5.24–5.26
"Zoning By-laws Map-landscapes, Toronto," capturing zoning "forms" for a downtown block in Toronto, inspired by Hugh Ferriss's staged drawings, using digital techniques.

restrictions in "S1" (Figure 5.24), followed in "S2" (Figure 5.25) by a more refined rendering that shows height restrictions and setback requirements. "S3" (Figure 5.26) is a further refinement of "S2," which includes sun angle penetrations and setback restrictions. Ferriss's maps/drawings contributed to an all-encompassing awareness of the effects of zoning and its architectural and urban design implications using the charcoal medium. Using digital processes, these map-landscapes express the urban container of this downtown section, based on zoning information. In his night scenes, Ferriss often illuminated building masses at their lower stories or at their summit, as if emitting an "aura," and often streaks of light in the sky highlighted the presence of the structure.[12] Lighting and shadow effects set the mood for the subject matter. Special lighting is placed at the bases and interiors of the forms to highlight the air and light corridors for these imaginary envelopes. In "S1," the whole mass is illuminated since it is primarily depicting height restrictions. In "S2" and "S3," extra omni lights are placed at the summit and lower stories to highlight changes in setbacks and sun angle requirements. Together, they can be used as sets of instruments that instruct the urban designer:

> The shadowy squares and translucent rectangles in *Zoning By-law*, "S1" could be from Hugh Ferriss, the great American architectural artist whose shadow-filled works from the early part of the 20th century haunt a lot of what Amoroso does. Comments from the general public include, "these images remind me of a ghostly city. – It looks like the conception of a new city."[13]

Light-switch maps

Selecting an appropriate and suggestive metaphor to depict the type of data is critical in manifesting the spatial or dynamic complexities of the city. For example, consider the entertainment district of the city. How can one map the "hot spots" or "happening" places in this area? Perhaps, one may consider a light-switch analogy, in which the buildings "flick" on with a bright light. This indicates an "awakening" of the building – it is ready for another day – and the light remains on until the building "shuts down" for the day. As more people enter these establishments, the buildings begin to glow brighter, indicating more intense use or occupancy (Figure 5.27 depicts the level of activity at 11.00 a.m. and Figure 5.28 represents levels of activity at 11.00 p.m.). These snapshot images are taken from the animated map depicting user occupancy over a 24-hour cycle. One can see the stronger levels of activity occur in the evening hours. Please see www.nadiaamoroso.com for the animated map depicting typical 24-hour activity levels (in buildings). You will also find another animation, *Hours of Operation, Gay District, Toronto*, that depicts the Church Street Spine. Various colors are applied to indicate specific uses. Red glows indicate convenience and grocery stores; green represents offices and banks; orange deals with cafés and fast food eateries; blue portrays restaurants, and white/yellow indicates

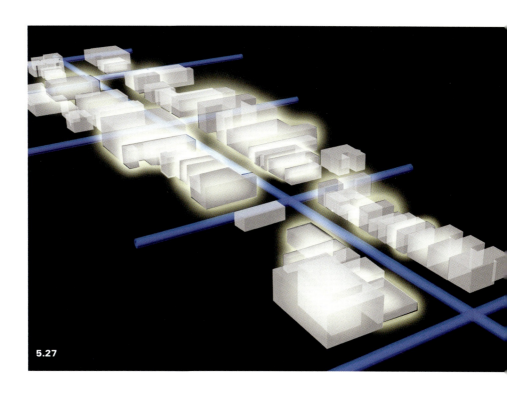

5.27

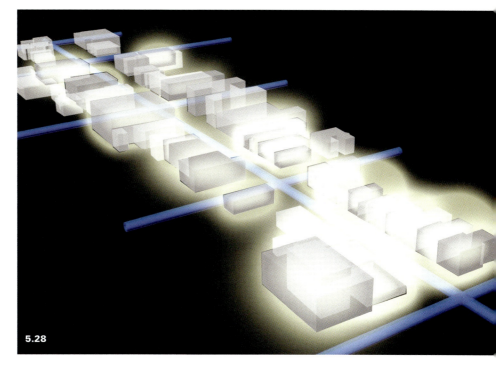

5.27–5.28
"Light-switch maps, Toronto," capturing "hot spots" or "happening" places in the Gay District. Using a light-switch analogy, the buildings "flick" on with a bright light, as the buildings house more people.

5.28

bars and clubs. When viewing this video clip, one begins to draw conclusions about the site. Cafés and certain eateries are lit up constantly from early until late. It is interesting to see a powerful stream of white-yellowish glow emanating from bars and clubs. Further, one begins to see a small red glow that is always "turned on" to indicate 24-hour convenience/grocery stores. The viewers are able to quickly distinguish pockets of activity at certain times of the day without specific searching for the store hours of operation.

Sculptural maps

The remaining illustrations in this chapter are a visual collection of sculptural maps crafted by graduate students interested in perceiving specific data types. Many of the spatial maps created use data type as a metaphor to generate the form or landscape. They are first conceived through a digital process, using data as the threshold of vertical and horizontal scales, in order to set the spatial limits and parameters of the form. In many instances, the data are dramatized, but nonetheless still hold true to the quantitative value. Students have achieved this by using certain methods of artistic license, including placing dramatic lighting at points of interest, using architectural and landscape architectural form-making techniques, and applying appropriate colors or textures. Some have taken the data and totally transformed them into radial architectural forms. In each case, however, a careful balance between the logic (information) and artistic (creative) is achieved. Finally, many of the students crafted the digital multidimensional maps as artifacts – either through computer numerical control or three-dimensional printing processes – transforming their maps as sculptural pieces and/or as models of potential urban forms.[14]

Rawlings Green, Cornell University

Figures 5.29 and 5.30 are by Joe Kubik (a former student of Nadia Amoroso at Cornell University). They depict "sources" and "sinks" of pedestrian movement surrounding the Rawlings Green site located on Cornell University campus. Points with positive elevations indicate the "sources" or areas generating pedestrian activity. The areas receiving pedestrian flow are assigned negative elevation values ("sinks"). Metaphorically, pedestrian flow is like water in which the movement runs from the higher to lower elevations. In essence, the Rawlings Green datascape is similar to a watershed, and transforms pedestrian movement into runoff. The viewer can begin to see that the trenches indicate the student pathways or circulation routes – carved out by the pedestrian movement. The holes in the sculptural map indicate large social gathering spots or deposit points. In essence, the sheer weight of the students creates holes in the landscape, indicating a direct relationship between weight (number of students) and the depth of the hole.

5.29

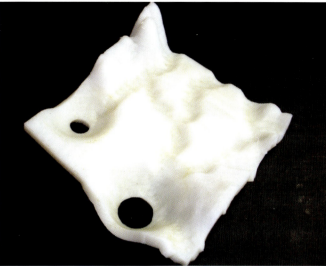
5.30

Bio-quad section, Cornell University

Figures 5.31–5.33 are by Starling Childs (a former student of Nadia Amoroso at Cornell University). The first two are map-landscapes for the Bio-quad section, located at Cornell University. The map, entitled "Flowscape," is a visual and spatial interpretation of the site's permeability, using measurements from a storm-water runoff as the data source. Consecutive data were mapped in the order of least permeable surfaces to most permeable. When mapping permeability across the site, specific preference was given to surfaces characteristic of a very low absorption rate such as rooftops and paved surfaces versus those elements of the site intended to mitigate runoff waters. The mapmaker in this case was interested to showcase permeability as a three-dimensional object. For example, a storm drain or swale creates a distinct dip in the model, while paved surfaces create positive extrusions. The idea that permeability would not be consistent along the entire length of a sidewalk is illustrated in the model by an undulation in the extrusions. These particular gestures relate to spot elevations on the sidewalk and where runoff waters are more likely to drain to. The digital model is created in a sensuous form and has a water-like texture, which indicates the subject of water. The plastic sculpture (Figure 5.32) portrays a smooth hill-like landscape that successfully captures the essence of "flowing" in a solid form.

5.31

5.32

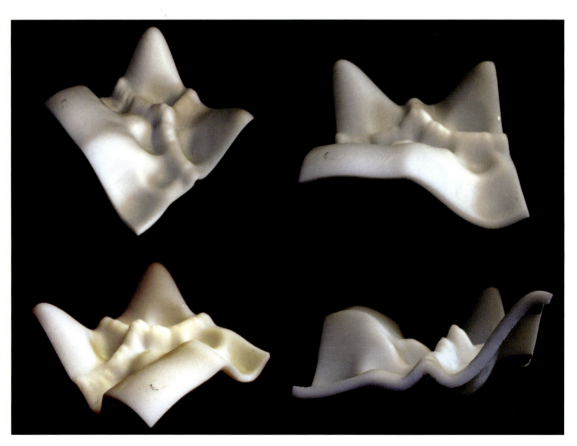

5.33

Chinatown, Toronto

The sculptural maps in Figure 5.34 are by Gavin Berman (a former student of Nadia Amoroso at the University of Toronto). They capture alternative ways to represent wind data statistics. Using a 1 kilometer by 1 kilometer base map for a specific location in Chinatown, Toronto, summer and winter wind pattern data for this area are transformed into spatial and sculptural maps.

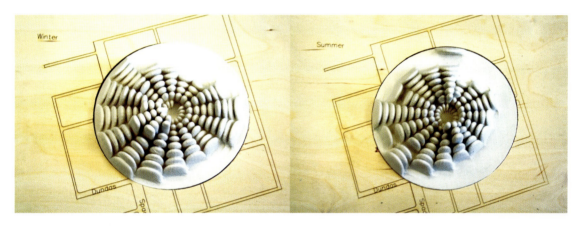

5.34

Little Italy, Toronto

The map-landscapes in Figures 5.35–5.37 are by Patrick Rudisuli (a former student of Nadia Amoroso at the University of Toronto). They craft the "hot spots" located in Little Italy, Toronto. Data relating to people movement within this specific area (at given times) are transformed into three sculptural artifacts. One map captures activity around the noon hour, the second captures people movement from early evening, and the final map expresses this data for the hour from 10.00 p.m. to 11.00 p.m. on the same weekday.

Distillery District, Toronto

The Distillery District in Toronto is known for its history of alcohol production and its currency as a trendy hotspot. The data set containing a survey of the alcohol-serving establishments of the area, a matrix of prices and types of alcohol served (Figure 5.38) was created by Robert Shostak (a former student of Nadia Amoroso at the University of Toronto). The field was then narrowed to bottles of wine served. Robert's sculptural maps represent this subset of data. A "Voronoi" diagram was generated for the space, creating boundaries around the specific establishments relating to proximity to each other (Figure 5.39). These boundaries inform the surfaces that represent the low prices of red and white wine on the bottom, followed by the high price of white wine through the middle, and the high price of red wine at the top. The surfaces are vertically supported by columns representing the number

Drawings

5.35

5.36

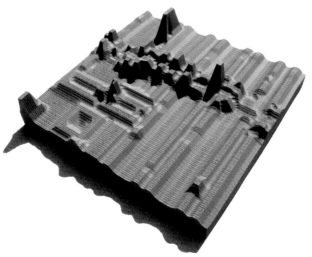

5.37

5.38

5.39

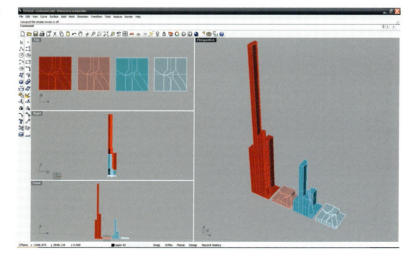

5.40

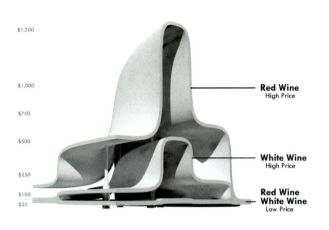
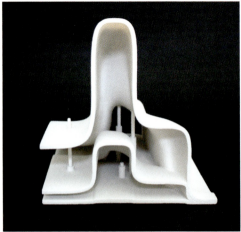

5.41   **5.42**

of different types of bottles offered by each restaurant (Figure 5.40) using Rhinoceros software. The thicker and lower bars are white wine while the taller, thinner bars are red wine.

Robert's final map resembles a model building design (Figure 5.41), and a physical model is generated – transforming the map as artifact – a synthetic figuration of the data (Figure 5.42). The bars are centered on the actual location of the restaurant within the Distillery District.

Casa Loma site, Toronto

Figures 5.43–5.45 are by Jesse Klimitz (a former student of Nadia Amoroso at the University of Toronto). The first captures the history of building development within the historical Casa Loma site in Toronto. As time progresses the peaks grow to illustrate the age of structures. The highest peak indicated in the map-artifact indicates the oldest building on site to the current date (Figure 5.44). The digital figuration of this map-landscape is transformed into an artifact that objectifies the data (Figure 5.45).

Liberty Village area, Toronto

Figures 5.46–5.49 are by Kate Slotek (a former student of Nadia Amoroso at the University of Toronto). The maps (Figure 5.46) fabricate the urban development process over one hundred years in the Liberty Village area in Toronto. The data set maps the built form – as the singular constant within the rapidly shifting urban fabric. Liberty Village contains a number of different formal conditions: the site contains both new and old buildings at a variety of scales. The mapmaker captured those elements of the building collection that have been altered and re-appropriated through diverse

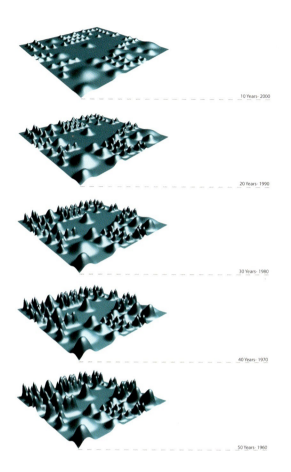
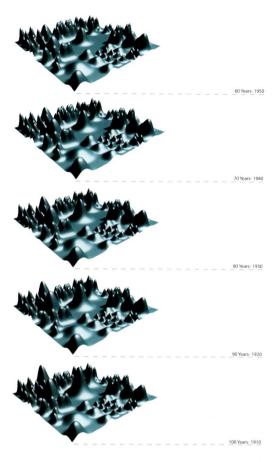

100 Years of Built History- Casa Loma

5.43

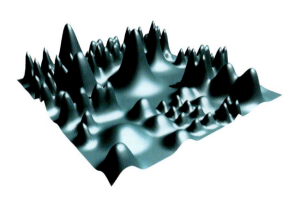

5.44

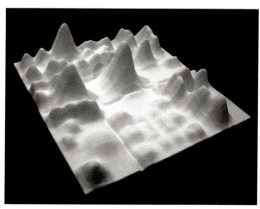

5.45

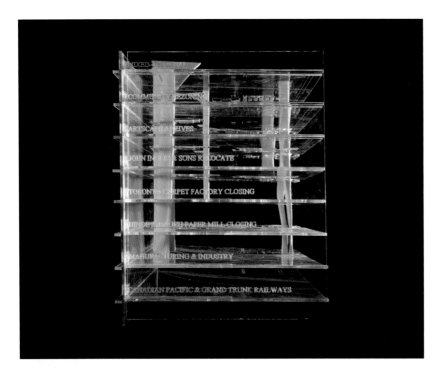

5.46

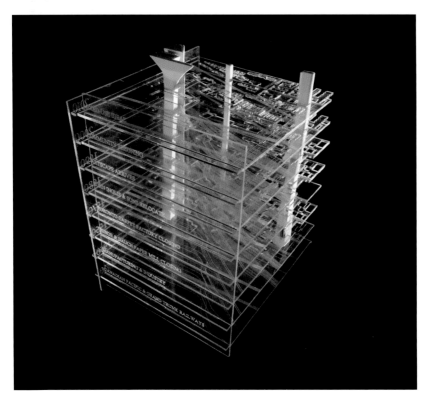

5.47

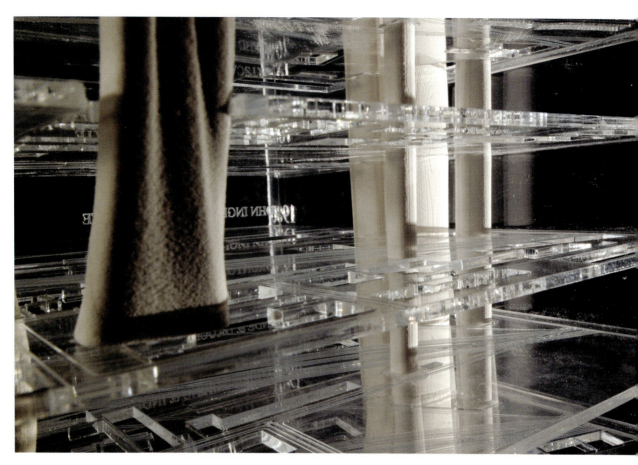

5.48

eras. The layered figure ground displays this change and evolution that has occurred through time; the bone white forms found within demonstrate endurance and continuity (Figures 5.47 and 5.48). Fabricated of a three-dimensional printed starch model, the form connects, twists, and intersects across the datum lines as seen in the digital rendition (Figure 5.49). This model was resolved through a diagramming exercise that traces change through the years; making visible the diverse processes of land parceling, subdivision, and zoning. In this way the built form, here temporal and abstracted, truly reflects the complexity of the urban condition. Once the industrial centre, now a chic hot spot, Liberty Village is a fascinating and unique neighborhood.

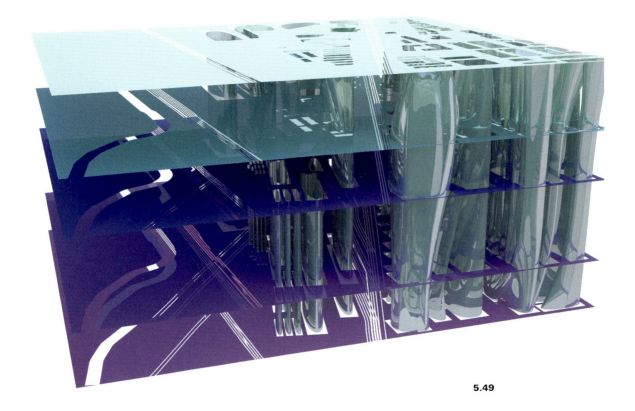

5.49

Final thoughts

The capacity of modern-day modes of representation to reveal unique interpretations of the city – to provide us with new and emotionally arresting bits of information – qualifies them as maps, though as maps quite distinct from any our parents or grandparents most likely would have seen. Revealing a multidimensional view of their subject, these new products are encoded with many layers of technical and abstract data reflecting the legal, environmental, economic, social and political circumstances within a city.

Through the manipulation of new drawing conventions and via the use of abstract signs, these new maps guide their viewers through the maze of an artificially constructed field of forces which define the physical reality of the built environment. However, these new maps can scarcely be viewed as purely "objective" representations: their artistic qualities and careful craftsmanship allow them to be subjective. They are, furthermore, highly controversial artifacts which register the prevailing social demands of their cultural context and also the personal input of the maker. As products of a thorough investigation of a wide range of environmental elements (invisible forces), these new maps produce a synthesis of the present and future conditions within a city.

These new maps become particularly revealing evidence of urban circumstances as they are perceived and felt by individual inhabitants. Described as simultaneously artistic and informative artifacts, these new maps reflect and graph both the complexity of the factors that define the present reality of an urban situation, and the processes which led to their creation. As sources through which predictions about the future of a city can be made, these maps serve as useful tools for urban planners and architects – in particular the animation clips which show changing experiences.

The works of Hugh Ferriss register as both artistic and informative images that qualify them as maps. He made these drawings over eighty years ago and they are still used as depiction reference for the 1916 New York City Ordinance. Ferriss's maps served as catalysts for the innovations which characterize the architectural movements of the early 1920s. His drawings/maps consolidated and provided a clear definition of the major legal and economic shifts within the milieu of early twentieth-century American culture which proved irresistible. In addition to all of this, Ferriss's images speak to all sets of audiences simply by virtue of their familiar architectural form, their style and their artistic quality.

A successful map-landscape should be *spatial, suggestive, seductive, informative, revelatory* – exposing the invisibles of the city through tantalizing new kinds of urban form – transforming banal data into a spatial outcome.

A summary of the guiding principles

A few guiding principles to create these map-landscapes include:

- *Treat data as spatial representations.* For example, take the number of people or zone index, and transform their numerical value as a spatial boundary. Tufte believes that one-dimensional data (such as population) are best represented through one-dimensional graphics. Two-dimensional data may be best represented by two-dimensional types of graphics. For example, where population is represented by circles, the area of each circle should be proportional to the population. The map-landscapes repurpose and transform all types of geo-referenced data into a new space. Therefore any dimensional data are represented through spatial forms. However, in creating map-landscapes in terms of exposing the hidden parameters of the city, it is essential to craft the "data value" as space or surface. The population density (the number of people per defined district) is proportional to the space (volume) created. The greater the number of people per defined area, the larger the mound (hill-form) is generated. As population numbers grow, so do the heights (peaks) of the city landscape, similar to the "Waste Sector" images produced by MVRDV. As architects, landscape architects and

urban designers, we are able to register peaks and valleys. This follows the next principle.
- *The visual representation of the data is related to the numerical representation.* Keep the volumetric portions directly related to the data value.
- *Use the data as the palette to guide the form.* Think of the data as space, and design the space based on the data restrictions.
- *Use effective artistic licenses.* Like Ferriss, who used his drawing skills to transform the textual confusions of the zoning laws as works of art and powerful guides into the unknown city.
- *Dramatize the data.* The data should be truthful but they have to be vivid and exciting. This will draw the attention of your audience. MVRDV dramatize data in an "unrealistic" way which nevertheless is true (not all the time) to the numerical facts, in particular as seen in the *Datatown* images. This is meant to shock the audience on the risks of continuing to consume, for example, high levels of energy or water. In the renditions of the *Crimescape* maps, they convey sharp and deep "pit-like" points that resemble an eerie, "hell-like" landscape. This is meant to draw attention to the districts in the city with high crime. The same can be said for the *Densityscape* maps. The original concept of the term "datascape" is an important factor, since it redefines information as "possible" space. In these cases, their *datascapes* achieved a balance between the aesthetic representation and the empirical presentation that can influence planning processes.
- *Choose an appropriate method of representation.* It should relate or respond to the data type, such as matching the data type to the design of the map-landscape – the data as a metaphor for the map-landscape. Wurman discussed the idea of "the map as a metaphor" – maps can provide people with the means to share the perceptions of others; they are the metaphoric means and the tools by which we can understand and act upon information from outside sources.[15] Spatial maps with a metaphorical reference suitable to the data at hand will convey the data in a more powerful way – the data as the metaphor. For example, the *Crimescape* maps are drawn with negative "pullings" to create deep depressions in the landscapes. The negative "pulls" are associated with the type of data – in this case, crime.
- *Apply more lighting to emphasize larger quantities or points of interest in the data.* In the surveillance space maps, for example, increased lighting is applied to "expose" those areas where surveillance cameras intersect or to emphasize areas of greater public surveillance in downtown Toronto. This rule is also applied in the crime maps, where extra lighting is positioned in the deeper pits representing higher crime readings. Ferriss clearly used the chiaroscuro effects to sensationalize the amount of light on the streets. This technique helps to convey his message of the importance of "returning" more sunlight onto the streets of New York.
- *Select the most telling viewpoint to profile the map-landscape.* Position your

map-landscape to achieve the optimum viewing outcome such as section-elevation, right-isometric, etc. Ferriss carefully positioned his subject matter to capture the most powerful view to convey the outcome of the data.

- *Visually represent the overall communicative message.* The maps should provide a strategic balance between the creative measure and the informative aspects of the data. The map-landscapes should guide the architect, landscape architect, urban planner and the common urban dweller to think more intuitively about their living space. For example, in the representation of crime rate indexes, a cognitive link is quickly drawn to the summits as the areas of high crime. The images allow their readers to draw critical insights about their cities that can shape development decisions and urban dynamics.

Tufte's life-long passion is a critical search for graphical excellence, which is a matter of substance, statistics and design, as the well-designed presentation of interesting data. In the works of MVRDV, datascapes serve as an architectural language, which turn statistical description into representations of the theoretical envelopes within which built forms can be placed. James Corner, specifically in his research on new modes of mapping, is unabashed in his concern about how "new things" can be represented as a map. The process of mapping and the means by which it is expressed visually are equally important. Like a design process, the mapping process is very exciting; the end result is often a surprise. But during the actual process, one discovers the "things" about the site (whether city or specific place) that scream out to be expressed in a visual form. Graphical and spatial inventiveness are just as important as the type of "obvious" urban information these representations expose. To put it another way, creativity can offer insight that straightforward tracing cannot. The connection between art and mapping information must never be downplayed; each – certainly the former – can add to the other. It is in the constant quest or creative process to deliver new visual means of expressing data that new discoveries are made about the urban spaces and places all around us.

Notes

1. Joseph R. Passonneau and Richard Saul Wurman. *Urban Atlas: 20 American Cities: A Communication Study Notating Selected Urban Data at a Scale of 1:48,000.* Cambridge, MA: MIT Press, 1966, p. 4. Please refer to this page in *Urban Atlas* for the complete propositions.
2. "Value system" means a scale value or set of values for areas or units such as grid squares, census tracts, etc.
3. These are grid lines which one can manipulate (i.e. push, pull, fold), similar to meshes or "chicken wire fencing" which have flexible wire grids.
4. Rodolphe el-Khoury is a professor and former director of the Urban Design Programme at the University of Toronto. El-Khoury provides this comment after viewing the exhibition on the Toronto maps at Lonsdale Gallery.
5. Two points about this. 1) It is not surprising that population densities should be higher

towards the centers of the cities compared with the periphery. 2) Whether the density is highest at the center of London depends on precisely where one defines the center to be. Of the above mentioned Boroughs, Islington and Camden are arguably the most central. The central Borough with low population is Westminster.

6   An artifact by definition is "1a: something created by humans usually for a practical purpose; especially an object remaining from a particular period (e.g. caves containing prehistoric artifacts). 2: a product of artificial character (as in a scientific test) due usually to extraneous (as human) agency" according to Merriam-Webster online dictionary. As discussed in a previous footnote, "artifacts" here refer to three-dimensional physical objects.

7   See the following website http://www.met.police.uk/crimefigures/. (2001 crime stats). When the mouse is placed over the district, a crime index pops up relating to the Borough. One can actually download this crime map.

8   London Crime Statistics, 2001. Westminster (8,483 total crime), Lambeth (7,904 total crime), Southwark (7,442 total crime) and Newham (7,344 total crime). http://neighbourhood.statistics.gov.uk/Default.asp?nsid=false&CE=True&SE=True (accessed June 2005).

9   A critique by Peter Goddard, a writer and art critic for *The Toronto Star* newspaper. This was his comment when he saw my new maps for Toronto showcased at a gallery.

10 Goddard's comment during his review of the work in *The Toronto Star* newspaper, 27 April 2002.

11 The locations in London included Bexley, Bloomsbury, Brent, London Bridge Place, Eltham, Hackney, Harlington, Hillingdon, Lewisham, Marylebone Road, North Kensington, Southwark, Sutton, Teddington, Wandsworth and Westminster.

12 Carol Willis, "Drawing towards Metropolis" in Hugh Ferriss's *The Metropolis of Tomorrow*, 1996, p. 155.

13 Peter Goddard, review of the work in *The Toronto Star*, 27 April 2002 (Art Review section).

14 The digital and sculptural maps created by the students at Cornell University were generated from the Landscape Architecture digital media elective course for graduate and undergraduate students. The digital and sculptural maps created by the University of Toronto students were part of the ARC 3033: Selected Topics in Architecture – Mapping Visualization course.

15 Wurman, *Information Anxiety2*, p. 157.

Afterword

Look Ahead Comments by SENSEable City Lab and Google Earth Inc.

Today there are a number of individuals and groups exploring alternative mapping techniques used as a means to "expose the invisibles" of the urban environment. Some tend to lean more towards scientific and informative measures of visualization, while others tend to push more the creative side and the art of mapping; yet others perhaps balance both aspects to reveal the new cities. Some even attempt to represent the invisibles of the urban realm using multidimensional mapping methods. Space Syntax Lab (The Bartlett School of Architecture and Graduate Studies, UCL), CHORA (UK), Christian Nold (UK), Brian McGrath (Manhattan TimeFormations), Spatial Information Design Lab (Columbia University), Eric Rodenbeck (Stamen Design, San Francisco), Nigel Holmes (mapping/ graphic explanation), Stephan Van Dam (NYC), Jack Dangermond (ESRI), Don Moyer (Thought Form Inc.), National Geographic Society, UNStudio (The Netherlands), Ole Bouman (NAI), Hans Rosling (Trendalyzer/Gapminder Foundation), Philip Beesley (University of Waterloo), Professor Perry Kulper, Architect (University of Michigan), Google Earth, SENSEable City Lab and the key contemporary key figures reviewed in this book are a handful of forward-thinking groups and individuals pushing the scope of visualization relating to exposing the invisibles of the city.

Hylozoic Soil

Hylozoic Soil is a lightweight expanded actuated diagrid of corrugated mesh integrated with underlying mechanisms that stir and pump humidified air. Figure A.1 shows *Hylozoic Soil* as seen from below and Figure A.2 as seen from the side. The mesh responds to human occupants with rolling swells of movement, forming a new "map-landscape." It is comprised of thousands of interlinking laser-cut elements, creating a "woven" topology of hyperbolic peaks and valleys. The particular topology expressed is a register of the forces within environmental systems acting all around them, and the precise positioning takes its form explicitly as a spatial map of those forces.

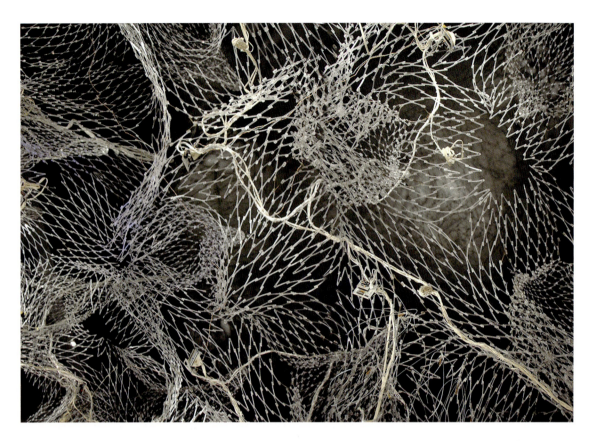

A.1 and A.2

Hylozoic meshwork canopy from *Hylozoic Soil* project by Philip Beesley with Rob Gorbet, Museum of Fine Art, Montreal, 2006.

Source: Philip Beesley Architect Inc. and collaborators

SENSEable City Lab and Google Earth

SENSEable City Lab (at MIT) and Google Earth are currently pushing visual mapping and GIS analysis into a new era – not much unlike the leap from McHargian paper maps, Corner's map-images, and the datascaping from MVRDV. Below are special features from SENSEable City Lab directed by Professor Carlo Ratti and from Mark Aubin, one of the co-founders of Google Earth.

SENSEable City: mapping the invisibles
(edited by Carlo Ratti, Nashid Nabian and Assaf Biderman, images by the SENSEable City Lab)

Mapping the invisibles in some spatial form is nothing foreign to a group of researchers at the SENSEable City Laboratory, directed by Carlo Ratti, at Massachusetts Institute of Technology. The works attempt to capture the everyday changes within our built environment. In many cases, the use of sensors and hand-held electronics like mobile phones or GPS has radically provided the group with the means to investigate the forces that affect the environment in some shape or form, and perhaps bring to light the theoretical new urban forms that these forces would generate.

One captivating image (Figure A.3) captures a mountainous landscape generated over the urban fabric of Rome. This new terrain is crafted from the data emitted by cellphone usage at a Madonna concert at the Olympic Stadium in Rome. As the

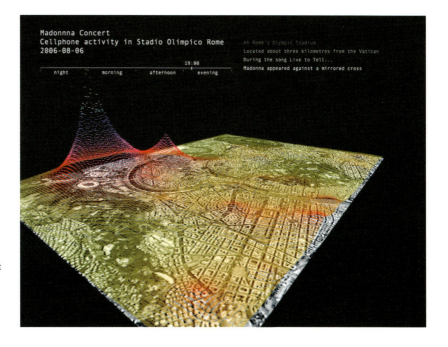

A.3
Real time cellphone activity at a Madonna concert in Stadio Olimpico, Rome.

Source: Provided by SENSEable City Lab, MIT

Afterword

concert is about to begin, a higher terrain is generated, indicating high cellphone usage during this time. A modest, more rolling hill-like landscape map is generated hours after the concert is finished, providing the viewer with an indication of the number of people left at this time.

In another project, entitled *currentcity.org*, a series of tantalizing "infoscapes" (as Ratti would call them) are generated by the aggregation of people who populate it. The SENSEable City, along with a Dutch Foundation (currentcity.org), uses the technologies and methods derived by the group to visualize a landscape or new theoretical city form based on public gatherings and traffic flux, and as such can visually detect which neighborhoods are the most crowded, and perhaps indicate alternative commuting patterns during the day. These new visuals, in essence, become new maps that guide the designer to accommodate other potential services for the public good, and for the city planner to think more critically about, for example, emergency planning, traffic management, the efficient allocation of utilities, or the impact of new city infrastructures. They have made the current city real, by leveraging those location data that our mobile phones generate over the course of the day. They have made the *currentcity* anonymous, aggregated and processed through innovative algorithms. Without invading on the privacy of individual subscribers to the network, their visualizations give important information on agglomeration and relative weights of human activities in the urban environment.

The new map-landscapes try to express data such as the number of people in the area, and perhaps help guide crowd management. The visuals also capture current demand for public transportation, and patterns of inflow and outflow of the

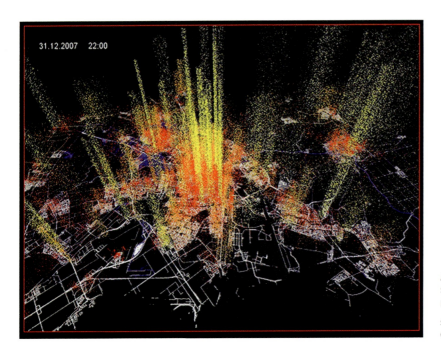

A.4
SMS activity at 10.00 p.m., New Year's Eve in Amsterdam.

Source: Provided by SENSEable City Lab, MIT

A.5
SMS activity at midnight, New Year's Eve in Amsterdam.

Source: Provided by SENSEable City Lab, MIT

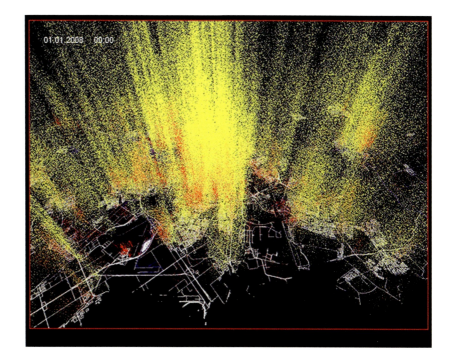

people in the city. These become additional guides for the urban planner to forecast better urban design decisions. The *infoscapes* can also generate data, based on the hot spots. The reader is able to view the social effects in the city and render the entertainment spots. These visualizations can also capture the number of people looking at a billboard, which in turn, can assist in marketing and city advertisement purposes. Figure A.4 depicts levels of SMS (texting) activities at 10.00 p.m. on New Year's Eve in a public space in Amsterdam. Figure A.5 visualizes the explosive level of SMS activity at midnight at the same place.

These visuals bring to the forefront information that was once unseen, and in turn, provide urban planners with better and more intuitive tools for wiser planning decisions. The works are about the responsive city, a city that reacts to the engaging citizens who occupy it. The works capture these responses in fascinating new forms of the city – a means to understand layers of data and phenomena.

Google Earth: the global mapping search tool
(edited by Mark Aubin, co-founder of Google Earth)

More and more individuals in the design and planning professions – urban designers, architects and landscape architects and of course students in these fields – are using Google Earth as the standard tool for visualizing, sharing and searching geographic information. In a conversation with Mark Aubin, one of the founders[1] of Google Earth, we discuss the ongoing popularity of Google Earth, as a "standard" for site searches, as a possible design tool and as a guide to help uncover, and perhaps "expose the hidden factors" of the city. Prior to using Google Earth, many landscape architects and urban planners relied on other mapping agencies and libraries to obtain aerial photos and site base maps. Now more than ever, capturing the site from the air is a few clicks away on Google Earth (but not all aerial imagery is high-quality).

From the early days at Keyhole Inc. to the now mega-Google Earth, Google Earth has become one of the most popular site-context search mapping platforms, with more than 500 unique user activations. In essence, Google Earth has become the de facto standard authority in the mapping realm. Due to its popularity, emphasis on high-quality representations and accurate portrayals of sites with their surrounding context are *a must* for Google Earth to stay competitive. One of the main purposes of Google Earth is to place sites into context with the rest of the whole geo-picture – *seeing complete areas – as a digital and public source.*

How has Google Earth helped expose certain invisibles of the city? One key feature available from its early days is a standard set of layers, including various boundaries such as school districts, postal codes, parks, cities, states and countries. For example, landscape architects are able to turn on the parks layer and see the boundaries that otherwise might not be visible, and take these into consideration before imposing any design interventions (Figure A.6). School district boundaries is another layer which allows urban planners and designers to investigate potential urban growth development around schools, and it also provides additional information to assist in understanding the market values for certain areas.

With the addition of "SketchUp," urban designers and architects can now place their building designs into a three-dimensional geo-context. The "3D buildings"

A.6
Park boundary mapping (New York City).

Source: © 2009 Google

layer in Google Earth is an incredible tool for design professionals concerned with the built environment, and is populated with models, many of which were built with SketchUp by Google Earth users. This layer provides a readily available mapping visualization tool to assist with design and planning decisions. Design students, architects, urban designers and landscape architects can "impose" their site or building design proposals directly into a three-dimensional geo-context. They are then able to evaluate their design proposals within the context of the existing landscape, providing a more realistic visualization of their potential changes.

The "historical imagery" feature introduced in version 5.0 gives users access to the past by providing a time slider that allows users to "go back in time." Urban designers and planners are able to interpret developmental growth patterns and changes in the landscape, now enabled through this new feature.

Aubin discusses one of the main transformations from the early days of Google Earth to today. When Google released Google Earth as a free consumer product, anyone with access to the web finally had access to imagery and rich mapping layers of the earth. While Google Earth was initially considered a viewing platform for geo-spatial content, it has become much more. This free access to viewing satellite, aerial, and mapping data in such an easy-to-navigate, seamless interface has opened people's eyes to their world. All sorts of professional fields now use Google Earth to enhance their business and daily experiences. Google Earth is a platform for people to place their journeys, research, and make discoveries in a geographical context. Individuals can also publish their own stories by creating KML documents that can be viewed in Google Earth. These KML documents allow users to save their own annotations on a map, such as route data, location comments, geo-referenced photographs, and recorded tours through the imagery. With these personal additions, an individual can visually publish their "local and personal stories" and provide a "sense of discovery" on this global platform.

In one incident, architects and planners were able to showcase their design proposals of a new section of the Bay Bridge in San Francisco. Using a SketchUp model of the bridge proposal placed into Google Earth, the architects were able to show the public the visual and contextual impact this new section of the bridge would have in the area (Figure A.7). During construction the architects have also been able to portray the "before and after" conditions of the new section of the Bay Bridge and see the progression of development. This visualization technique guides designers into making wiser decisions.

In the summer of 2008, Mark Aubin joined the Google Earth Outreach team on a trip to Brazil where they led workshops for the indigenous people of the Amazon rainforest on how to use technology to help protect their lands and preserve their cultures. Mapping is a key component of protecting their lands because it allows them to accurately document their historical borders, hunting grounds, sacred sites and cultural sites. Publishing their maps on Google Earth allows indigenous people to tell their own story to the world.

Afterword

A.7
3D modelling and mapping using Google SketchUp in Google Earth (Bay Bridge, San Francisco).

Source: © 2009 Google

In another case, public health officials used Google Earth to locate the invisibles of swine flu incidents. By mapping individual cases, they were able to visualize the distribution and spread of the virus.

The future of Google Earth is to increase the coverage of the data and the options available for viewing. Layers such as 'street view' show eye-level street conditions of particular sites. More and more cities will be added as the data are gathered. Google Earth is also trying to provide a more complete collection of readily available historical maps, and more accurate visual portraits of the earth. Google Earth provides a fascinating way to expose the invisibles of the city and bring them to the visual forefront.

A.8
The "Densityscape" image of Manhattan's 2001 population density, geo-positioned into Google Earth.

Source: © 2007 Google/Nadia Amoroso

A.9
A snapshot moment of the "Market-valuescape" of Manhattan, geo-referenced into Google Earth.

Source: © 2007 Google/Nadia Amoroso

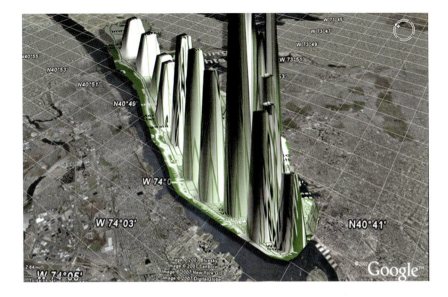

A.10
A snapshot moment of the traffic volume at 8.00 p.m. on a particular day in New York City, geo-referenced into Google Earth.

Source: © 2007 Google/Nadia Amoroso

Google Earth can potentially become the global platform in which creative and spatial maps can be viewed at an ever-changing level – to show the factors such as change in populations (Figure A.8), changes in the city's market level (Figure A.9) or traffic flux (Figure A.10) or whatever types of urban forces that tend to stay hidden yet shape the city's landscape.

Note

1   Other co-founders now at Google include John Hanke, VP and Michael Jones, Chief Technology Advocate. They were some of the key individuals of Keyhole Inc., which was acquired by Google in 2004, and whose product became Google Earth.

Bibliography

Abrams, Janet and Peter Hall, eds. *Else/Where: Mapping New Cartographies of Networks and Territories*. Minnesota: University of Minnesota Design Institute, 2006.

Adams, Ann Marie and Pieter Siljkes. "Architects are Heroes at Metropolis; Exhibit Features All-Star Guest List of 20th-Century Designers." *The Gazette* (Montreal, Quebec) 13 July 1991: J2.

Adams, Thomas. "The Building of New York City." *Regional Plan, vol. 2. New York: Regional Plan of New York and its Environs*, 1931.

Amery, Colin. "The Potent Vision of Hugh Ferriss: Architecture." *Financial Times* (London) 12 Oct. 1987: 21.

Amoroso, Nadia. "Mapping the Invisible." *Canadian Architect*, January 2003, p. 38.

—— "Invisible Urban Forms: Quantitative Data as Inspiration for Design." *Archis,* No. 2, May/June 2003, pp. 98–101.

—— "Redrawing the Map." *The Next American City,* Issue 8, Spring 2005, pp. 44–45.

—— "The Exposed City: Inspiration for New Urban Form." *Urban Design* (London), Issue 98, Spring 2006, pp. 16–17.

Banerjee, Tridib and Michael Southworth, eds. *City Sense and City Design: Writings and Projects of Kevin Lynch*. Cambridge, MA: MIT Press, 1990.

Boyd, John Taylor Jr. "The New York Zoning Resolution and its Influence upon Design." *Architectural Record*, 3, September 1920, pp. 193–217.

Bunschoten, Raoul. "CHORA: Proto-Urban Conditions and Urban Change." *Architectural Design* (London), vol. 66, January/February 1996, pp. 16–21.

—— *CHORA/Raoul Bunschoten: From Matter to Metaspace: Cave, Ground, Horizon, Wind*. New York: Springer, 2005.

—— and Helene Binet. *Urban Flotsam: Stirring the City*. Rotterdam: 010 Publishers, 2001.

—— *CHORA*. 2007. http://www.chora.org/.

Burchard, John and Bush-Brown, Albert. *The Architecture of America: A Social and Cultural History*. Boston: Atlantic Monthly Press; Little, Brown Books, 1961.

Center for History and New Media, George Mason University. "Maps: What Makes a Map a Map?" 2003–2005. http://chnm.gmu.edu/worldhistorysources/unpacking/mapswhatmakes.html.

—— "Maps: Why Bother with Maps?" 2003–2005. http://chnm.gmu.edu/worldhistorysources/unpacking/mapswhybother.html.

Ciolkowska, Muriel. "Hugh Ferriss and the Zoning Laws of New York." *Architectural Review*, November 1925, pp. 173–177.

Clark, W. C. and Kingston, J. L. *The Skyscraper: A Study in the Economic Height of Modern Office Buildings*. Cleveland, New York: American Institute of Steel Construction Inc., 1930.

Clark University. *On Mapping Projects*. 2005. http://www.clarku.edu/offices/publicaffairs/news/press/2005/mapping.cfm.

Confurius, Gerrit. "Editorial." *Daidalos,* 69/70, 1998, pp. 7–10.

Corbett, Harvey W. "High Buildings on Narrow Streets." *The American Architect,* 2369, June 1921, pp. 603–619.

—— "Zoning and the Envelope of the Building." *Pencil Points,* 4, April 1923, pp. 14–16.

—— "The Planning of Office Buildings." *Architectural Record,* 41, September 1924, p. 92.

—— "New Stones for Old." Three-part series. *Saturday Evening Post,* 198, 27 March, 8 May, 15 May 1926.

—— "The Skyscraper: Babel or Boon?" *The New York Times* 5 Dec. 1926: SM1.

—— "New York in 1999 – Five Predictions: Architects and City Planners Look into the Crystal Ball and Tell What They See." *The New York Times,* 6 Feb. 1949: SM18.

Corner, James. "A Landscape at Work." *The Chronicle of Higher Education,* 43, 15, December 6, 1996, p. B96.

—— "Operational Eidetics: Forging New Landscapes." *Harvard Design Magazine,* Fall 1998, pp. 22–26.

—— "Eidetic Operations and New Landscapes." *Recovering Landscape: Essays in Contemporary Landscape Architecture.* Ed. James Corner. New York: Princeton Architectural Press, 1999, pp. 152–169.

—— ed. *Recovering Landscape: Essays in Contemporary Landscape Architecture.* New York: Princeton Architectural Press, 1999.

—— "The Agency of Mapping: Speculation, Critique and Invention." *Mappings.* Ed. Denis Cosgrove. London: Reaktion Books, 1999, pp. 213–252.

—— and Alex MacLean. *Taking Measures across the American Landscape.* New Haven, CT: Yale University Press, 1996.

Cosgrove, Denis. "Carto-City." *Else/Where: Mapping – New Cartographies of Networks and Territories.* Eds. Janet Abrams and Peter Hall. University of Minnesota Press: University of Minnesota Design Institute, 2006, pp. 148–165.

—— ed. *Mappings.* London: Reaktion Books, 1999.

Czerniak, Julia. "Challenging the Pictorial: Recent Landscape Practice." *Assemblage,* 34, December 1997, pp. 110–120.

—— ed. *CASE: Downsview Park Toronto.* London: Prestel, 2001.

Danzer, Gerald. "Introduction: Analyzing Maps." (lecture) Center for History and New Media, George Mason University, 2003–2005. http://chnm.gmu.edu/worldhistorysources/analyzing/maps/analyzingmapsintronf.html.

Davenport, Thomas H. and Laurence Prusak. *Information Ecology: Mastering the Information and Knowledge Environment.* New York: Oxford University Press, 1997.

Duffus, R. L. "The Metropolis of Tomorrow: Mr. Ferriss considers the Problem created by the Skyscraper." *The New York Times,* 8 Dec. 1929: BR6.

Dunlap, David. "Commercial Real Estate: Turning Radiator Building into a Boutique Hotel." *The New York Times,* 11 Aug. 1999: 6.

—— "The Design Image vs. the Reality." *The New York Times,* 28 Sept. 2003: 1.

Embury II, Aymar. "New York's New Architecture: The Effect of the Zoning Law on High Buildings." *The Architectural Forum,* 4, October 1921, pp. 119–124.

Emery, H. G. and K. G. Brewston, eds. *The Century Dictionary of the English Language.* New York: The Century Co., 1927.

Ferriss, Hugh. "Civic Architecture of the Immediate Future." *Arts and Decoration,* November 1922, pp. 12–13.

—— "The New Architecture." *The New York Times,* 19 March 1922: 52–54.

—— "Truth in Architectural Rendering." *The Journal of the American Institute of Architects,* 13, March 1925, pp. 99–101.

—— "The Skyscraper Climbs to a New Height." *The New York Times,* 5 Sept. 1926: SM3.

—— "Rendering, Architectural." *The Encyclopedia Britannica,* 14th Edition, 1929–1973 (revised in 1961 and titled as "Architectural Rendering").

―― "How Hugh Ferriss Draws: Six Progressive Stages in Rendering a Proposed Building." *American Architect*, July 1931, pp. 30–33.
―― *The Metropolis of Tomorrow*. Princeton, NJ : Princeton Architectural Press, 1986.
―― *The Metropolis of Tomorrow*. Mineolta, NY: Dover Publications Inc., 2005.
Forgey, Benjamin. "The Poet of the Skyline; Hugh Ferriss and his Projections from the 1920s." *The Washington Post*, 28 Feb. 1987: G1.
Frye, Curtis D. *Metacity/Datatown* [book review], 2000. http://www.techsoc.com/metacity.htm.
Girot, Christophe. "Four Trace Concepts in Landscape Architecture." *Recovering Landscape: Essays in Contemporary Landscape Architecture*. Ed. James Corner. New York: Princeton Architectural Press, 1999, pp. 58–67.
Goldberger, Paul. *The Skyscraper*. New York: Knopf, 1981.
―― "Architecture: Renderings of Skyscrapers by Ferriss." *The New York Times*, 24 June 1986: 13.
Hastings, Thomas. "The Zoning Regulations in New York." *The American Architect*, 2338, October 1920, pp. 461–463.
Harley, J. B. and Woodward, David. *The History of Cartography*. Chicago: The University of Chicago Press, 1987.
Harney, Andy Leon. "Signage: Fitting Letter Forms to Building Form." *AIA Journal*, October 1975, pp. 35–39.
Hastings, Thomas. "The Zoning Regulations in New York." *Architectural Forum*, Oct. 1921, p. 125.
Hecht, Heiko, Robert Schwartz and Margaret Atherton, eds. *Looking into Pictures: An Interdisciplinary Approach to Pictorial Space*. Cambridge, MA: MIT Press, 2003.
Hume, Christopher. "Unfulfilled Promises of the '20s Hits Home at Metropolis Exhibit." *The Toronto Star*, 22 June 1991: F1.
Huxtable, Ada Louise. "Four Centuries of Drawings – a Record of Vision and Taste." (Architecture View) *The New York Times*, 27 Apr. 1980: 31–32.
Iovine, Julie V. "Dutch Designs for Cities Built on Ideas and What-If's." *The New York Times*, 15 October 2002: p. 1.
Jakle, John A. "Reviewed work(s): *Taking Measures across the American Landscape* by James Corner; Alex S. MacLean." *Annals of the Association of American Geographers*, 87 (3) Sept. 1997, pp. 538–539.
Jensen, Bill. *Simplicity: The New Competitive Advantage in a World of More, Better, Faster*. Cambridge, MA: Perseus Publishing, 2000.
Jewell, Edward A. "America's Power Portrayed in Art: Drawings of Great Buildings by Hugh Ferriss Shown at the Whitney Museum." *The New York Times*, 5 May 1942: 16.
Johns, Orrick. "Architects Dream of a Pinnacle City." *The New York Times*, 28 Dec. 1924: SM10.
―― "The Excelsior of Architecture." *The New York Times*, 20 July 1924: SM3.
―― "Our Billion-Dollar Building Year." *The New York Times*, 14 Sept. 1924: SM7.
―― "Bridge Homes – A New Vision of the City." *The New York Times*, 22 Feb. 1925: SM5.
Kant, Immanuel. "Book II: Analytic of the Sublime – The Mathematically Sublime." *The Critique of Judgement*. Oxford: The Clarendon Press, original print 1790, reprint 1964, section 25.
Kapsenberg, Jan. "Erotische Manöver + ein Vorschlag für die Umwandlung des Thermalbades in Vals von Peter Zumthor." *Daidalos*, 69/70, 1998, pp. 76–81.
Kilham, Walter H. *Raymond Hood, Architect: Form through Function in the American Skyscraper*. New York: Architectural Book Publishing Inc., 1973.
Leich, Jean Ferriss. *Architectural Visions: the Drawings of Hugh Ferriss*. New York: Whitney Library of Design, 1980.
Litt, Steven. "Glorious Renewal Looms for Shabby Police Building." *Plain Dealer* (Cleveland, Ohio), 7 Mar. 1993: 3H.
Lootsma, Bart. "Towards a Reflexive Architecture." *El Croquis: MVRDV*, Vol. 86, 1997, pp. 34–43.
―― "July 1997 (on datascapes)." *Berlage Cahiers* 7, p. 39.
―― "Synthetic Regionalization: The Dutch Landscape Toward a Second Modernity."

Recovering Landscape: Essays in Contemporary Landscape Architecture. Ed. James Corner, 1999, pp. 250–274.

—— "Biomorphic Intelligence and Landscape Urbanism." *Topos*, 40, 2002, pp. 12–20.

—— "The Diagram Debate, or the Schizoid Architect." *Datascaping*. 2002. http://predmet.arh.unilj.si/siwinds/s2/u4/su4/S2_U4_SU4_P6.htm.

—— "What is (really) to be Done?" *Reading MVRDV*. Ed. Veronique Patteeuw. Rotterdam: NAi Publishers, 2003. pp.24–63.

Lutticken, Sven. "MVRDV: Stroom HCBK – The Hague – Architectural Firm Exhibition." *Art Forum*, December 2001. http://findarticles.com/p/articles/mi_m0268/is_/ai_80856220.

Lynch, Kevin. *The Image of the City*. Cambridge, MA: MIT Press, 1960.

—— *A Theory of Good City Form*. Cambridge, MA: MIT Press, 1982.

Maas, Winy. *Metacity/Datatown*. Rotterdam: MVRDV/010 Publishers, 1999.

—— *KM3: Excursions on Capacities*. Barcelona: Actar, 2005.

MacAdam, George. "Vision of New York That May Be: A Forecast of Manhattan Transfigured by the Zoning Law." *The New York Times*, 25 May 1924: SM2.

Makielski, S. J. Jr. *The Politics of Zoning: The New York Experience*. London and New York: Columbia University Press, 1966.

McGuigan, Cathleen. "How Yesterday Saw Tomorrow." *Newsweek* (US edition), 19 Aug. 1991: 52.

Mostafavi, Moshen and Ciro Najle, eds. *Landscape Urbanism: A Manual for the Machinic Landscape*. London: Architectural Association, 2003.

Mumford, Lewis. "The Sacred City." *The New Republic*, 27 Jan. 1926, 270–271.

—— "Magnified Impotence." *The New Republic*, 22 Dec. 1926: 138–140.

The New York Times. Hugh Ferriss's drawing for the advertisement "Does It Pay to Look into the Future? What Will Happen to New York Real Estate in the Next Hundred Years?" *The New York Times*, 8 Mar. 1923: 10.

—— "Architects League Honors Hugh Ferriss." *The New York Times*, 2 May 1941: 15.

—— "Art Notes." *The New York Times*, 1 May 1942: 24.

—— "Architects in City Choose New President at Meeting." *The New York Times*, 4 June 1942: 14.

—— "Ferriss Heads Architects: Portrayer of City of the Future Installed by League Here." *The New York Times*, 7 May 1943: 16.

—— "Consultant is Elected Head of Architects." *The New York Times*, 5 June 1952: 54.

—— "Hugh Ferriss, 72, Architect Here; Farseeing Designer Is Dead – Foe of Skyscrapers." *The New York Times*, 30 Jan. 1962: 29.

The Nonist. "Hugh Ferris: Delineator of Gotham." *The Nonist: Hugh Ferris*, 2003. http://thenonist.com/index.php/thenonist/permalink/hugh_ferriss_delineator_of_gotham/ (accessed 22 May 2008).

Passonneau, Joseph R. and Richard Saul Wurman. *Urban Atlas: 20 American Cities; a Communication Study Noting Selected Urban Data at a Scale of 1: 48,000*. Cambridge, MA: MIT Press, 1966.

Patteeuw, Veronique (ed.) *Reading MVRDV*. Rotterdam: NAi Publishers, 2003.

Pond, Irving K. "Zoning and the Architecture of High Buildings." *The Architectural Forum*, 4, October 1921, pp. 131–134.

Ratti, Carlo and Nick Baker. "Urban Infoscapes: New Tools to Inform City Design and Planning." *ARQ (Architectural Research Quarterly)*, 7, 1, 2003, pp. 63–74.

Real Estate Record Association. *A History of Real Estate Building and Architecture in New York City During the Last Quarter of the Century*. New York: Arno Press, 1967.

Robinson, Cervin and Bletter, Rosemarie H. *Skyscraper Style: Art Deco, New York*. New York: Oxford University Press, 1975.

Sadler, Simon. *The Situationist City*. Cambridge, MA: MIT Press, 1998.

Sanders, James. "Art/Architecture: Taking the Memorial Designs for a Test Drive." *The New York Times*, 30 Nov. 2003: 40.

Schumacher, Patrick. "The Dialectic of the Pragmatic and the Aesthetic: Remarks on the Aesthetic of Datascapes." (lecture at Architectural Association, London), 1997. http://www.patrikschumacher.com/aesthetics.htm.

Shane, Grahame D. "The Emergence of 'Landscape Urbanism': Reflections of *Stalking Detroit*." *Harvard Design Magazine: On Landscape,* Fall/Winter 2003, pp. 1–8.

—— "The Emergence of Landscape Urbanism." *The Landscape Urbanism Reader*. Ed. Charles Waldheim. New York: Princeton Architectural Press, 2006. pp. 55–67.

Silver, Mike and Diana Balmori. *Mapping in the Age of Digital Media.* London: Wiley-Academy, 2003.

—— "Images/Matter." *Else/Where: Mapping – New Cartographies of Networks and Territories*. Eds. Janet Abrams and Peter Hall. University of Minnesota Press, 2006, pp. 206–211.

Skupin, Andre and Katy Borner. "Mapping Humanity's Knowledge and Expertise in the Digital Domain." *Environment and Planning B: Planning and Design 2007,* 34, pp. 765–766.

Smith, R. E. *Psychology*. New York: West Publishing Company, 1993.

Sommer, Robert. "A New Mural Movement Brings Indigenous Art to the Streets." *AIA Journal,* October 1975, pp. 32–33.

Soukhanov, Anne and Dr. Kathy Rooney, eds. *Encarta Webster's Dictionary of the English Language*. Second Edition. London: Bloomsbury, 2004.

Steele, Brett. "Reality Bytes: Datascapes." *Daidalos,* 69/70, 1998, p. x.

Steindorff, Ulrich. "Architecture of Power – Three Periods." *The New York Times*, 6 July 1924: SM3.

Stern, Robert *et al. New York 1930: Architecture and Urbanism between Two World Wars.* New York: Rizzoli, 1987.

Sullivan, Chris C. "Model Data (Technology)." *Architecture*, 91, 11, 2002, p. 26.

Sulsters, Willem A. "Mental Mapping, Viewing the Urban Landscapes of the Mind." (Conference paper) in www. Tudelft.nl.

Swan, Herbert S. "Making the New York Zoning Ordinance Better: A Program of Improvement." *The Architectural Forum*, 4, October 1921, pp. 125–130.

Toll, Seymour I. *Zoned American*. New York: Grossman, 1969.

Tufte, Edward R. *Envisioning Information.* Cheshire, CT: Graphics Press, 1990.

—— *Visual Explanation: Images and Quantities, Evidence and Narrative.* Cheshire, CT: Graphics Press, 1997.

—— *The Visual Display of Quantitative Information.* Second edition. Cheshire, CT: Graphics Press, 2001.

UK Government. *Crime Stats, London, UK.* http://www.met.police.uk/crimefigures.

UK Government. *London Crime Statistics, 2001.* http://neighbourhood.statistics.gov.uk/Default.asp?nsid=false&CE=True&SE=True.

Van Koov, Mathew. "The Authority of the Architect: Concerning Discourse and Method." (Online paper, Architecture + Philosophy public lecture series 2, Federation Square, Melbourne), April 2007, pp. 1–13. www.architecturephilosophy.rmit.edu.au.

Waldheim, Charles. "Aerial Representation and the Recovery of Landscape." *Recovering Landscape: Essays in Contemporary Landscape Architecture*. Ed. James Corner. New York: Princeton Architectural Press, 1999, pp. 120–139.

Wall, Alex. "Programming the Urban Surface." *Recovering Landscape: Essays in Contemporary Landscape Architecture*. Ed. James Corner. New York: Princeton Architectural Press, 1999, pp. 232–249.

Ward, David and Olivier Zunz, eds. *The Landscape of Modernity: Essays on New York City, 1900–1940.* New York: Russell Sage Foundation, 1992.

Weinberg, George. "Dutch Firm Envisions City of the Future." *The Yale Herald*, 6 Sept. 2002: 2.

Weller, Richard. "An Art of Instrumentality: Thinking Through Landscape Urbanism." *The Landscape Urbanism Reader*. Ed. Charles Waldheim. New York: Princeton Architectural Press, 2006, pp. 69–85.

Whitney, William Dwight. *The Century Dictionary*. New York: Century, 1911.

Willis, Carol. "Zoning and Zeitgeist: The Skyscraper City in the 1920s." *Journal of the Society of Architectural Historians,* March 1986. pp. 47–59.

—— "The Titan City: New Building Technologies and Theories of Urban Planning in the 1920s Gave Rise to Intoxicating Visions of Future Cities." http://www.americanheritage.com/articles/magazine/it/1986/2/1986_2_44.shtml.

—— "Unparalleled Perspectives: The Drawings by Hugh Ferriss." *Daidalus*, 15 September 1987, No. 25, pp. 78–91.

—— "A 3-D CBD: How the 1916 Zoning Law Shaped Manhattan's Central Business District." *Planning and Zoning New York City: Yesterday, Today and Tomorrow.* Ed. Todd W. Bressi. New Brunswick: Center for Urban Policy Research, Rutgers University, 1993.

—— *Form Follows Finance: Skyscrapers and Skylines in New York and Chicago*. New York: Princeton Architectural Press, 1995.

Wood, Denis. *The Power of Maps*. New York: Guilford Press, 1992.

Wurman, Richard Saul. *Making the City Observable*. Cambridge, MA: MIT Press, 1971.

—— *Cities: Comparisons of Form and Scale*. Philadelphia, PA: Joshua Press, 1974.

—— "An Interview with the Commissioner of Curiosity and Imagination of the City of Could Be." *AIA Journal,* April, 1976, pp. 62–63.

—— *Information Anxiety: What to Do When Information Doesn't Tell You What You Need to Know*. New York: Doubleday, 1989.

—— *Rome Access*. Fourth Edition. New York: Access Press, 1995.

—— *Information Architects: What to Do When Information Doesn't Tell You What You Need to Know*. New York: Graphis Inc., 1997.

—— *Information Anxiety2*. Indianapolis, IN: Que, 2001.

—— *Richard Saul Wurman's UnderstandingUSA*. 2006. http://www.understandingusa.com/wurman.html.

—— *19.20.21*. Richard Saul Wurman website, 2008. www.192021.org.

Zmudzinska-Nowak, Magdalena. "Searching for Legible City Form: Kevin Lynch's Theory in Contemporary Perspective." *Journal of Urban Technology*, 10, 3, 2003, pp. 19–39.

Index

abstraction xiv, 49, 56, 85, 100–2, 106, 118, 135–6
Access travel guidebooks 52, 55–6, 58, 66–7
aesthetics 21, 37, 39, 47, 84, 106, 117
agency 99, 105, 109, 112; agent 5, 99, 105–6
air-quality index (map-landscape) 140
analogue 100
analogy 137, 142
apocalyptic approach 88
"architectural delineator" 5, 24, 25
artifact xii–i, 3, 24, 30, 85, 125, 144, 147, 150, 154–5, 158
artistic viii, xi–iii,xv, 3, 5, 6, 15, 23, 26, 32, 34, 53, 68, 82–7, 94–5, 97–8, 101–2, 106, 109, 117, 119, 123, 129, 140, 144, 154–6
attraction; attractive 22, 23, 55, 86, 102
aura 7, 142

"bombastic" 84
Boullée 4, 35
boundary 53, 72, 125, 137, 155, 164
Brunier, Yves 108
bulk 8, 11–12, 18, 20, 31, 140
bureaucratic 69–70

"chartjunk" 61, 63, 95

chiaroscuro 7, 8, 24, 119, 156
citymaker 86
"communicative architecture" 90
communicative force xiii, 35
computer numerical control (CNC) 85, 125, 133, 144
computerization 63
consumption 32, 68, 71, 72, 73, 76, 86, 87
contextuality 100
conventional xi, 14, 29, 50, 69, 83, 97
Cooper, Muriel viii
Corbett, Harvey W. 8, 9, 10, 11, 16, 19, 32
credibility 6, 19, 20, 25, 34, 94
crimescape 125, 129, 156
"cubical chaos" 22
currentcity.org 162
"cyborgian" 84
Czerniak, Julia 101

Daily News Building 20, 125
Danzer, Gerald 100
datascape viii, xii, xv, 64, 68–73, 76–7, 80, 82, 84–90, 144, 156–7
Deleuze, Gilles 110
densityscape 121, 123, 125, 156, 166
"directiveness" 90

distortion 87, 94
dramatize; dramatization 25, 72, 87, 88, 89, 125, 144, 156
dynamic 26, 60, 63, 87, 112, 117–18, 121, 137, 142, 157
dystopian 72, 82, 85

ecology 106
eidetic 69, 99, 100–1, 108
Eisenman, Peter 110
emotional response 48
empirical viii, 61, 117, 156
ephemeral 29, 109
Ernst, Max 109
Evolution of the Set-back Building drawings; "Stage 1" drawing, 7, 8, 140; "Stage 2" drawing, 12, 13, 14; "Stage 3" drawing, 13, 14; "Stage 4" drawing, 14, 15, 16; final stage, 10
extrapolation 86
"extreme scenarios" 71, 73

fantasy 72–3, 83
FARMAX 89
field xiv, xv, 3, 42, 45, 46, 49, 59, 70, 73, 84, 136, 147, 154, 163, 165
five rings of information 54
flexible bands 123
Ford, George 6, 11, 12, 23